Sustainable Construction

This book focuses on the utilization of wastes produced from plastic usage by industry and consumers, along with the partial to full replacement of conventional cement as a primary binder material in concrete. This book demonstrates how to use post-consumer waste plastics and industry wastes from thermal power stations, agro-industries, and metal industries with a scientific approach to conventional concrete. The primary aim is to demonstrate the methods to prepare a sustainable alternative construction material of concrete using waste materials.

Features:

- Illustrates making eco-friendly procedures of concrete construction popular by way of utilization of plastics and industrial wastes
- Covers all major aspects of plastic waste-based concrete from conception to execution
- Promotes alternative materials for sustainable construction
- Describes economic aspects of using eco-efficient concrete on a mass scale
- Includes experimental results with graphs

This book is aimed at researchers and graduate students in civil engineering, construction materials, and concrete.

Sustainable Construction
Development of Eco-Efficient Concrete with Plastic and Industrial Wastes

Ankur C. Bhogayata

CRC Press
Taylor & Francis Group
Boca Raton London New York

CRC Press is an imprint of the
Taylor & Francis Group, an **informa** business

Designed cover image: © pixbay

First edition published 2024
by CRC Press
6000 Broken Sound Parkway NW, Suite 300, Boca Raton, FL 33487-2742

and by CRC Press
4 Park Square, Milton Park, Abingdon, Oxon, OX14 4RN

CRC Press is an imprint of Taylor & Francis Group, LLC

ISBN: 9781032527956 (hbk)
ISBN: 9781032621333 (pbk)
ISBN: 9781032621340 (ebk)

DOI: 10.1201/9781032621340

Typeset in Times
by codeMantra

Contents

About the Author

Ankur C. Bhogayata is an Professor in the Department of Civil Engineering, Marwadi University, Rajkot, Gujarat, India. He has 20 years of work experience including teaching and research activities. He obtained his PhD from Saurashtra University in Civil Engineering with specialization in Concrete Technology. He has been granted with one patent and three design patents by the Intellectual Property Rights, Patent Office of Government of India. He has published more than 16 research articles in SCI and SCOPUS index journals with reputed publishers such as Elsevier, Springer, and ACI journals and has earned 11 h-index and 11 i10-index with more than 431 citations. His areas of expertise include concrete composites, geopolymer concrete, precast concrete, application of AI-based optimization methods for mix design of concrete, sustainable concrete, use of nano-particles in concrete, and study of structures exposed to marine environments. Presently, more than six research scholars are pursuing PhD under his supervision and one student obtained his PhD in the academic year 2021. He is the recognized reviewer for reputed journals also. He has been awarded with IGIP recognition by the International Society of Engineering Pedagogy, Austria, for his innovative teaching-learning practices for the students in Architecture and Civil Engineering programs. He has supervised more than 24 Master's degree students for thesis and also guided several undergraduate students for their research projects in the field of concrete technology. He has received funded projects from SSIP, NewGen IEDC, and GUJCOST for research work and academic activities. Music, painting, and reading are his hobbies. He is a member of the Institution of Engineers India, ISTE, and a life member of the India Chapter of American Concrete Institute.

Preface

Sustainability is key to the survival of humanity on mother earth. Nearly all small and large activities we do on a daily basis consume energy, resources, and materials from nature. It is the right time to observe *Civil Engineering Construction* as one of the most consistently growing endeavors from time immemorial from a sustainability perspective for the 21st century. There is a need to reduce the use of conventional natural resources in today's fast-developing construction. The most common and obvious choice of construction material is cement-based concrete. However, concrete has remained the second-largest consumed material by humans after water on the earth, which results in environmental pollution, loss of natural resources, and waste generation in natural streams. An alternative construction material to reduce and replace the conventional concrete produced with ordinary cement and ingredients, namely natural aggregates and sand, has become an essential requirement.

This book focuses on the utilization of wastes produced from plastic usage by industry and consumers along with the partial to full replacement of conventional cement as a primary binder material in concrete. This book demonstrates how to use post-consumer waste plastics and industry wastes from thermal power stations, agro-industries, and metal industries with a scientific approach. The primary aim is to demonstrate the methods to prepare a sustainable construction material as an alternative of cement-based concrete using waste materials.

This book explains basic methods for processing, treatments, and mix designs for preparing concrete using wastes obtained from post-consumer waste plastics and industrial wastes, namely fly ash, slags, and silica fumes mixed with alkaline solutions. This book includes experimental studies, results, analysis, and conclusions retrieved from the laboratory investigations on materials prepared using different waste materials. The strength and durability properties are assessed and discussed in the modified concrete. The tests on fresh state and hardened state concrete have been presented explicitly to understand the role and impact of various wastes on the behavior of the modified concrete. Suggested applications of the modified concrete are also included. The environmental impact and sustainability aspects of conventional and modified concrete have been discussed in detail in the appropriate sections of this book. The information in the book is expected to fulfill scientific and technological requirements of researchers, students, and professionals. The data, results, and discussion may serve as state-of-the-art knowledge for beginners and advanced learners in the field of sustainable concrete development.

Acknowledgments

This book is a collection and narration of the technical and scientifically obtained data on the methods and ways of transforming conventional concrete into a sustainable material. To serve the purpose, the experimental work has been carried out on the materials prepared with concrete consisting of plastic wastes, fly ash, silica fume, and furnace slag in different compositions. This book is a work performed by the team of students, colleagues, and laboratory technicians in the Department of Civil Engineering at Marwadi University, Rajkot, Gujarat, India.

The author is thankful to Marwadi University, Honorable Provost Prof. (Dr.) Sandeep Sancheti, Dean Research Dr. R. B. Jadeja, and a close friend and colleague Dr. Tarak Vora for their support and help. The author expresses special thanks to Dr. Kaushik Rao, Librarian, Marwadi University for his help and coordination with CRC press as and when required.

The author extends his gratitude to Dr. Gagandeep Singh – Senior Publisher (STEM) of CRC Press, Taylor & Francis Group, for supporting with all the communications, clarifications, and unconditional support required from the author at nearly each stage of this book. It may be said that this book is the outcome of the efforts of the author and encouragement of Dr. Singh at CRC Press. The author would like to express his gratitude to Dr. N. K. Arora, Professor and Head, Department of Civil Engineering, Government Engineering College, Gandhinagar for being the source of inspiration and showering his blessings being a Guide and mentor. In addition, the author gratefully acknowledges the support from Mr. Abhay Nakum and Mr. Shemal Dave for being sincere students and contributors in the laboratory investigations.

The author sincerely acknowledges all the support of each staff who have remained directly and indirectly involved in the experimental work.

1 Introduction

1.1 NEED FOR THE DEVELOPMENT OF ECO-EFFICIENT CONCRETE

General

Mankind has been utilizing enormous natural resources for its self-development and progress-related activities. However, it is difficult to enlist all of them; some of the most significantly used natural materials are water, stone, fossil fuels, sand, and timber. If we restrict the discussion related to construction activities, all these materials are extensively in use since many centuries. The interesting thing is that even with the significant and revolutionary changes in the trends of construction and building practices from the stone age till today, there are natural materials being constantly used and with more demand than in earlier times.

We all know that most of the natural materials are limited as they have been generated by nature over a period of millions of years and are difficult for us to reproduce. This fact should help humans to maintain a "balance" in nature as we are the only species on earth capable of extracting useful materials from natural sources and unfortunately never been able to pay back to nature for its loss caused by us.

Emerging from this deep-rooted philosophy of resource utilization and restoration, we, being civil engineers, should focus on the mitigation of the use of natural materials. Also, we have been polluting nature due to our construction- and building-related activities and that has become another huge challenge globally without any apparent solutions for management and safe disposal. This is exactly the background score for us to begin the discussion on the need for eco-efficient concrete.

1.1.1 SOME NOTES ON ECO-EFFICIENT CONCRETE

The question is, why does concrete require to be transformed into an eco-efficient concrete? To explore the possible answers, let us check the following aspects first:

a. Concrete: Making and Development Process Processing
   Concrete is a man-made composite for construction prepared with water, cement, aggregates, and sand; its preparation requires natural materials in a large quantity. Nearly at every stage of its processing, a large amount of energy is required, for which further we need fossil fuels. Moreover, processing of limestone into cement, crushing of rocks into its aggregates, and hauling of sand from the riverbeds have imposed irreversible loss of natural materials and generated pollution at various processes in the making of concrete. It is required to think to what extent such practices will last. Also, can we compensate the loss to nature? The answers are known to us; however, we do not pay serious attention due to several reasons. But now, at present,

DOI: 10.1201/9781032621340-1

we face nonavailability of some of these materials. For example, sand is one such material not available in many areas and parts of the world compared to earlier times. Many governments have declared a ban on using river sand in construction in many states and countries. This situation will deteriorate with time and will be the same for other materials also.

b. Concrete: Production and Energy Requirements

Concrete uses materials produced in nature by natural processes; however, we may not be able to use them unless they are pre-processed with bare minimum activities like excavation (mining), cleaning, sorting, burning, crushing, transportation, and storage. If we observe, all these activities need energy in large quantities. The unfortunate aspect is that for producing energy again we require natural materials! Hence the equation of the making of concrete shows only usage plus usage and no corresponding return on the other side. The energy demand, however, is difficult to specify for cement or concrete preparation; the average energy needs of this material may be a figure of 4–5 GJ of fuel for 1 ton of cement production (Worrell et al. 2008).

c. Concrete: Reuse and Waste Disposal

Concrete hardens with time and acquires strength eventually. However, being a permeable material, concrete shows deterioration and cracking of mass. The deterioration may be due to internal or external impurities and the environmental conditions for most of the cases. As a result, the functional life of concrete gets reduced and shows signs of moderate to severe failure. Such concrete or its components are subjected to demolition and become construction and demolition wastes. Such wastes are one of the huge issues of environmental pollution, especially due to land occupation. Therefore, concrete is nowadays being considered as non-environment-friendly material. A review has showed concerns regarding this situation (Menegaki et al. 2018). Nearly 35% of the waste concrete and other demolished wastes are sent to landfills. Undoubtedly, the figure is likely to increase with time. The waste concrete is largely explored as the replacement for natural aggregates called "recycled concrete aggregates" in several references. The application is challenged by excessive water absorption, poor compression resistance, and difficulties of bonding and internal coherence mechanism in the matrix (Lauritzen 1998). Though the reference is old, it presents important concerns that remain valid even now.

d. Concrete: Green Initiatives and Present Scenario

The term "green initiatives" refers to the concept of reuse and reduce of the concrete waste and natural ingredients. Nowadays, several practices are in place to mitigate the usage of natural materials in concrete. For example, efforts have been made to reduce the greenhouse gases emitted in the cement manufacturing process (Adesina et al. 2020). The waste concrete also has been explored in making self-compacting concrete (Sun et al. 2020; Malazdrewicz et al. 2023). These are a few examples of attempts done by researchers at the laboratory level for creating concrete using waste concrete and thereby to reduce the natural conventional concrete-making material and reduction in the challenges imposed by the waste concrete.

However, even at present, the waste created by concrete buildings has continued to generate pollution, and there has been increasing demand for natural materials than earlier. The following may be some of the factors affecting such unsuccessful scenarios on waste usage and addition of natural materials in concrete:

- The demolished or waste concrete requires pre-treatment like crushing and breaking, which requires energy; hence, it may not be claimed as energy-efficient use.
- The waste concrete consists of other wastes, namely rebars, wires, wall paint colors, or other impurities induced, namely corrosion and by-products of similar deterioration. Such things make the waste difficult to be recycled to a single entity, for example, an aggregate.
- The waste concrete is not measured for its strength parameters. This will impact the strength aspects of the new concrete being prepared. There is always an uncertainty associated with the waste concrete sources, and it may not be offering uniform behavior or response to the test conditions.
- In some cases, transporting the waste from the source to a central recycling facility may prove to be uneconomical.
- In summary, we may note that the reuse or recycling of concrete waste in making of green or sustainable concrete composite is not an effective or a feasible option to a certain extent. From the above discussion, there is now need for making an eco-efficient concrete by some means. The same is discussed in the next section.

1.2   RECENT TRENDS OF RESEARCH IN THE DEVELOPMENT OF ECO-EFFICIENT CONCRETE

We have discussed that preparing concrete using the waste concrete is difficult and full of challenges. However, the researchers have made significant efforts in making the concrete economically cheaper and environmentally safer by using various waste products. In this section, let us see the wastes that have been explored for their possible inclusion in concrete.

1.2.1   RECENT RESEARCH TRENDS IN ECO-EFFICIENT CONCRETE

a. Cement Alternatives in Concrete

Cement has remained the most energy-intensive material in concrete preparation and the primary material to generate the strength mechanism for the ingredients. Therefore, a careful replacement is required for cement in the concrete mix. Some recent references dealing with the partial replacement of cement with the other materials, primarily waste products, are discussed here. Rock dust has been added into the mixture of concrete as a partial replacement of cement by up to 15% by weight (Dobiszewska et al. 2023). The study showed reduction of strength and durability of cement upon increasing its replacement with rock dust. An observation showed that up to a 15% replacement, the dust worked as filler material and increased the surface area of the binder component including cement. The replacement did not adversely affect the strength; the researchers noted that excessive replacement of cement reduced the strength and durability of the concrete. Kumar et al. (2023) explored the possibilities of using grounded granulated blast slag and fly ash as the partial replacement of the cement in concrete. The results were encouraging, and the team found that the addition of such pozzolanic binder materials in blended mode with cement could enhance the workability and strength properties of the mixes. To further improve concrete's eco-efficiency, advanced mix-design methods must be applied; as a result, an overview of the various particle packing models (PPMs) is provided in a research work (De Grazia et al. 2023). The work emphasizes the value of adopting other materials to partially replace cement while focusing on the environmental issues around cement. To further improve concrete's eco-efficiency, cutting-edge mix-design methods must be applied. It is suggested to combine PPM with limestone fillers as an alternative ready-to-apply way to create eco-efficient combinations, enabling concrete to remain competitive as a sustainable material in the future. To obtain a pure form of pozzolanic binder material, a team of researchers worked on the pre-treatment of bottom ash (Abdulmatin et al. 2023). The aim was to obtain the chloride resistance for the concrete. The bottom ash was finely grounded and blended with the ground bottom ash of coarser particle size with the gross replacement of cement up to 25%. The team obtained increment of 35% in the strength and low penetrability of the chloride ingress.

There are several similar examples available to refer to and may be followed in future researches wherein the conventional cement may be replaced by the partial addition of other pozzolanic waste materials. The research work has shown that the eco-efficiency of the concrete depends on the replacement ratio and the strength criteria to be obtained. The strength of concrete is a function of the cement content of the mixture and therefore, the replacement and desired strength should be balanced adequately.

b. Sand Alternatives in Concrete

One of the essential natural ingredients of concrete is sand. Till date, river sand is being considered as one of the primary sources of sand for concrete. However, as sand alternatives, the crushed form of limestones and weathered stones is already being used in practice. Ultimately, all are natural resources and require attention for their rate of consumption. Researchers have made efforts to alter the sand use from a partial to full scale worldwide.

Pebble sand has been explored for possible addition in concrete by Srinivasan et al. (2023) recently. The aim was to use crushed pebbles in place of fine aggregates in the production of concrete. The prepared pebble sand replaced up to 50% of natural sand and the concrete was evaluated for different properties, namely strength in compression, and the modified concrete showed increment of the load resisting capacity in compression. Manufactured sand is one of the recent developments in the field. The fresh and hardened properties of concrete consisting of 100% manufactured sand have been assessed (Patel et al. 2023). The findings indicated that the addition of grounded granulated blast furnace (GGBFS) and manufactured fine (MF) sand decreased workability by about 25% while increasing mechanical strength by up to 50%. When GGBFS and MF were added, the SEM–EDS analysis revealed the presence of C–S–H and C–A–S–H gel in comparison to control concrete. The optimized concrete, which contains 50% cement, 30% GGBFS, and 20% micro fine, emits around 50% less CO_2 in GWP and costs 24% less than the control concrete. The work indicated eco-efficiency and sustainability of modern concrete without using natural sand. Use of recycled glass is another recent exploration regarding the partial replacement of sand in concrete (Mansour et al. 2023). Finer particle size results in increased surface area and improved intermolecular adhesion that resulted in excellent density of the mix. The marine dredged waste's fine particles were used in the concrete as sand replacement for up to 50% in the concrete (Hayek et al. 2023). Primarily the dredged material is the sediment material at the bottom of the sea and being a waste requires timely removal. Therefore, using waste materials in concrete also supports sustainability. Replacing up to 30% of the natural sand, the finer particles of the sediments did not adversely affect the concrete quality. Another research showed the utilization of eggshell in a crushed form for the partial replacement of sand in concrete (Hakeem et al. 2023). Interestingly, the strength in compression of the concrete consisting of 15% of the eggshell as fine aggregates increased due to the replacement. However, it was observed that 10% replacement of the sand proved to be efficient and provided adequate strength to the concrete. Several studies have been conducted on the use of agricultural wastes, namely rice husk ash, palm oil ash, bagasse ash, as the partial replacement of natural sand in concrete in varying proportions. The burning of these wastes creates pollution and requires a solution. On the other hand, use of such ash in a large quantity is important. Therefore, to satisfy both the requirements, the ashes of such waste agricultural products have attracted the attention of the researchers.

Stratoura et al. (2023) have used rice husk ash along with perlite as fine aggregates in concrete. The team found excellent improvement of the microstructural properties owing to the inclusion of the fine-sized ash as the partial replacement of the natural sand. The addition of the ash improved the chloride penetration, workability, and strength properties. The micropores of the concrete were found to be filled with the ash substances and improved the molecular adhesion by filler effects. The lightweight concrete has been developed using olive biomass ash in concrete (Cuenca-Moyano et al. 2023). The addition of the ash reduced the thermal conductivity of the conventional concrete by up to 45% and reduced the density when added up to 35% in the concrete mixes. Though the replacement was for cement, an alternative material known as expanded clay has also been studied; the utilization of the biomass supported the development of the sustainable concrete in general. In one of the interesting works, the use of sewage sludge has been experimented as fine sand replacement in concrete. The optimal replacement with 5% of the natural sand with sewage sludge sand showed excellent responses against the compressive strength and coherence of the material. The water absorption was also reduced by the hardened concrete mass consisting of innovative sludge sand.

Similarly, several agricultural, industrial, and other wastes have been used and explored by researchers to reduce the use of natural sand in concrete. It has been observed that with adequate pre-processing and waste treatment, such wastes may be used for field applications also. However, the study on the long-term durability of the modified concrete with alternative materials in concrete for sand replacement has remained an area of exploration to an extent.

c. Alternative Aggregates in Concrete

Aggregates like the crushed stones of 10–20 mm sizes contribute nearly 50%–60% of the total concrete volume. Therefore, the production of eco-efficient concrete can be more advantageous in case of replacement of the natural aggregates with waste-based aggregates. Like sand as the finer size stones, the aggregates with varying sizes from 10 to 20 mm have been replaced by alternative materials, largely the waste materials. Let us see some of the recent works on the topic.

Compared to cement and sand, the coarse aggregates have been replaced with a wider range of alternative materials. Coconut shells have been utilized as coarse aggregates for concrete (Bhoj et al. 2023). The natural aggregates were replaced up to 25% by coconut shells in the M30 mixture grade. The study revealed that only 15% of the replacement of the aggregates showed balanced response of the concrete and there was no adverse effect observed on the strength and other properties. The shell waste could be utilized with pre-processing and appropriate size rendering. The material may be useful where lightweight concrete is to be developed without preliminary concern of the strength properties. The combination of waste marble and basalt has been employed in another research (Boğa et al. 2023). Though the type of the concrete was self-compacting concrete, the properties for fresh and hardened concrete were assessed and the natural aggregates were replaced by up to 100%. Up to 25% of the total aggregate volume was replaced by marble waste and found useful for the enhancement of the properties. Nowadays, demolition waste from the old concrete structures is being explored for possible inclusion in the new concrete as aggregates. A review paper showed the strength and cost comparison of different recycled aggregates (Basit et al. 2023). In fact, the use of recycled demolished material in concrete should be analyzed for the cost factor along with the technical feasibility because the overhead expenses and logistics pertaining to such utilization play a significant role. Another research involving the blended formation of aggregates from crushed coconut shells and demolished brick residues may be referred as an innovative approach (Liu et al. 2023). The researchers have made efforts to have dual advantage of waste utilization in the preparation of concrete. Importantly, the combined use of such waste improved the thermal conductivity, cost ratio, and flexure response of the modified concrete.

The material became light weight and may be utilized for the structures with seismic importance. Though the compressive strength showed a small decline in the values, the concrete showed good response for the lightly loaded applications. The ceramic industry is another potential source of waste resulting from the processes of tile manufacturing. A research team (Zhang et al. 2023) developed concrete with fine aggregates prepared from waste tiles and observed an increment of 15% and 25% in the compressive and flexure strength of the concrete, respectively. Though the addition of tile wastes reduced the flow characteristics of the fresh mixtures, the strength values showed an improvement.

There are few other recent developments around the use of various wastes in concrete in the form of aggregates of fine and coarse sizes and available for future studies. Summarizing, it may be noticed that the eco-efficiency of the concrete modified with the waste materials in varying forms should be checked appropriately for strength and durability aspects. Based on the review and references available on the use of wastes in concrete, the following general guidelines may be extracted:

- The preliminary properties of the waste, namely chemical composition, physical properties and characteristics, thermal and other mechanical properties, should be thoroughly checked before use.
- The quantity or the amount of replacement of the waste in concrete should be systematically and scientifically investigated by performing the necessary tests for fresh and hardened properties. The optimum dosage is an important aspect for waste in concrete.
- Overall cost or cost comparison of the modified concrete with the conventional concrete should be carried out. This is important because though the waste utilization is a preliminary objective of making concrete sustainable, the economic aspects should be checked well for making the modified concrete feasible in the field applications.
- Along with the short-term evaluation of the modified concrete containing wastes, the long-term durability assessment should be made clear by holding time- or age-dependent tests on the specimens. This is again important as the service span of the concrete structures is nearly 30–50 years in most parts of the world. However, maintaining consistent service and integrity of concrete has a significant importance for its quality. Therefore, the wastes being utilized in the concrete and the modified concrete mixes should be cross-checked for its long-term durability and responses to varying loading conditions.

The readers are advised to refer to authentic and scientifically developed documents and literature on the topic of waste utilization in concrete before beginning with their studies. This is necessary because the author has observed that there are many waste materials providing good or excellent results at the preliminary stage of investigations; however, the durability and long-term behavior do not support the modifications. In such cases, the support of standards and codes may be referred along with the material development process with the waste inclusion. The next section summarizes such attributes along with the noticeable outcomes on the waste utilization in concrete toward making eco-efficient concrete.

1.3 SUMMARY OF SIGNIFICANT OUTCOMES AND AREAS FOR RESEARCH IMPROVEMENT

This section is a summary of the overall achievements of researchers in the development of sustainable eco-efficient concrete from wastes. The literature and the database suggest following points to be included in the summary:

- Industrial wastes may be utilized in three forms, namely powder, fine, and coarse-sized aggregates in conventional concrete.
- Conventional cement may be replaced with powder-like wastes such as fly ash, silica fumes, and furnace slag without significant pre-treatments. The optimum replacement ratio should be less than 50% to maintain the minimum binder characteristics of the matrix.
- The natural sand may be replaced by fine-sized particles rendered from industrial and agricultural as well as community wastes. However, the particle size, preliminary properties, and the composition of the wastes must be evaluated before use. There is no upper cap for replacement ratio observed and the complete replacement may be possibly included in the modified concrete.
- The natural coarse aggregates may be replaced by a variety of aggregates such as construction wastes, plastic wastes, glass waste, agricultural waste, and ceramic wastes. Practically, the list may be more than this. However, the strength criteria should be checked well in advance by performing pilot studies. The packing density, intermolecular interaction, and the load transfer patterns should be studied in detail before the use of waste aggregates in conventional concrete.

With these introductory remarks and preamble to the main topic of the book, let us explore the ways to improve the sustainability aspects of conventional concrete using some specific wastes, namely plastic waste and industrial waste. The book focuses on the utilization of non-biodegradable plastic wastes in conventional concrete in various ways by replacing the natural materials and on the complete replacement of conventional cement from concrete by utilizing the industrial wastes and chemicals resulting in one of the most advantageous binder materials commonly referred to as geopolymer-based concrete or alkali-activated concrete.

REFERENCES

Abdulmatin, Akkadath, et al. "Bottom ash as an alternative pozzolanic material to produce eco-friendly, high-quality chloride-resistant concrete." *Iranian Journal of Science and Technology, Transactions of Civil Engineering* 47.2 (2023): 829–842.

Adesina, Adeyemi. "Recent advances in the concrete industry to reduce its carbon dioxide emissions." *Environmental Challenges* 1 (2020): 100004.

Basit, Md Abdul, et al. "Strength and cost analysis of concrete made from three different recycled coarse aggregates." *Materials Today: Proceedings* (2023). (Article in press) https://doi.org/10.1016/j.matpr.2023.03.247

Bhoj, Siddhartha, A. Manoj, and S. Bhaskar. "Usage potential and benefits of processed coconut shells in concrete as coarse aggregates." *Materials Today: Proceedings* (2023). (Article in Press) https://doi.org/10.1016/j.matpr.2023.03.529

Boğa, Ahmet Raif, and Ahmet Ferdi Şenol. "The effect of waste marble and basalt aggregates on the fresh and hardened properties of high strength self-compacting concrete." *Construction and Building Materials* 363 (2023): 129715.

Cuenca-Moyano, Gloria M., et al. "Design of lightweight concrete with olive biomass bottom ash for use in buildings." *Journal of Building Engineering* 69 (2023): 106289.

De Grazia, M. T., L. F. M. Sanchez, and A. Yahia. "Towards the design of eco-efficient concrete mixtures: An overview." *Journal of Cleaner Production* 389 (2023): 135752.

Dobiszewska, Magdalena, et al. "Utilization of rock dust as cement replacement in cement composites: An alternative approach to sustainable mortar and concrete productions." *Journal of Building Engineering* 69 (2023): 106180.

Hakeem, Ibrahim Y., et al. "Eggshell as a fine aggregate replacer with silica fume and fly ash addition in concrete: A sustainable approach." Case Studies in Construction Materials 18 (2023): e01842.

Hayek, Mahmoud, et al. "Valorization of uncontaminated dredged marine sediment through sand substitution in marine grade concrete." *European Journal of Environmental and Civil Engineering* (2023): 1–18. https://doi.org/10.1080/19648189.2023.2168765

Kumar, Anant, and Krishna Deep. "Experimental investigation of concrete with cementitious waste material such as GGBS & fly ash over conventional concrete." *Materials Today: Proceedings* 74 (2023): 953–961.

Lauritzen, E. K. "The global challenge of recycled concrete." *Sustainable Construction: Use of Recycled Concrete Aggregate: Proceedings of the International Symposium organised by the Concrete Technology Unit, University of Dundee and held at the Department of Trade and Industry Conference Centre. London, UK on 11–12 November 1998.* Thomas Telford Publishing, 1998.

Liu, Haibao, Qiuyi Li, and Peihan Wang. "Assessment of the engineering properties and economic advantage of recycled aggregate concrete developed from waste clay bricks and coconut shells." *Journal of Building Engineering* 68 (2023): 106071.

Malazdrewicz, Seweryn, Krzysztof Adam Ostrowski, and Łukasz Sadowski. "Self-compacting concrete with recycled coarse aggregates from concrete construction and demolition waste-current state-of-the art and perspectives." *Construction and Building Materials* 370 (2023): 130702.

Mansour, Mohammed A., et al. "A systematic review of the concrete durability incorporating recycled glass." *Sustainability* 15.4 (2023): 3568.

Menegaki, Maria, and Dimitris Damigos. "A review on current situation and challenges of construction and demolition waste management." *Current Opinion in Green and Sustainable Chemistry* 13 (2018): 8–15.

Patel, Dhaval, et al. "Experimental investigation on mechanical properties of ternary blended concrete using manufactured sand." *Innovative Infrastructure Solutions* 8.5 (2023): 136.

Srinivasan, G., and J. Saravanan. "Pebble sand (P-sand) as a partial replacement for fine aggregate in concrete." *IUP Journal of Structural Engineering* 16.1 (2023): 28–40.

Stratoura, Maria C., et al. "Perlite and rice husk ash re-use as fine aggregates in lightweight aggregate structural concrete-durability assessment." *Sustainability* 15.5 (2023): 4217.

Sun, Chang, et al. "Utilization of waste concrete recycling materials in self-compacting concrete." *Resources, Conservation and Recycling* 161 (2020): 104930.

Worrell, Ernst, Christina Galitsky, and L. Price. *"Energy Efficiency Improvement and Cost Saving Opportunities for Cement Making."* LBNL-54036-Revision. Ernest Orlando Lawrence Berkeley National Laboratory, University of California, March (2008).

Zhang, Liqing, et al. "Effect of ceramic waste tile as a fine aggregate on the mechanical properties of low-carbon ultrahigh performance concrete." *Construction and Building Materials* 370 (2023): 130595.

2 Plastic Wastes

2.1 TYPES OF PLASTICS AND THEIR USAGES FOR CONSUMER-ORIENTED ACTIVITIES

2.1.1 BACKGROUND

Post-consumer plastic waste contributes about 10%–19% of municipal solid waste and about 43% of total plastic waste. More than 400 million tons of plastic are produced annually worldwide. In the last 60 years plastics have been utilized in large expanses of industrial, commercial, and household activities and have become an intricate part of our daily life. Plastics are suitable to be used repeatedly by recycling, making them economically viable material for many industries including packaging. However, overuse of plastic has resulted in hazardous environmental impacts due to its low degradation property. Hence, satisfactory management and safe disposal of plastic waste has become a global challenge for all of us. Plastics are non-biodegradable materials and exist in nature for a prolonged time, adversely affecting land, water, and air. Innovative methods of plastic waste disposal are urgently needed to mitigate its hazardous impacts on the environment.

The major proportion of plastic production is consumed by packaging activities followed by construction use. The usage scenario reflects the need for innovative methods for safe disposal of the waste generated by post-consumer activities. Plastic usage has changed and replaced many conventional materials. The application ranges from a normal carry bag to aircraft. Plastic could be the most revolutionary material ever invented by human beings. It holds the capacity to fit into a wide range of products that are useful in daily life. Plastic is present in countless many forms around us and probably shall remain as an ever-changing material. One of the major advantages of plastic is its flexible form and capability to attain desired shape and strength for a long service life.

Considering the quantities produced as non-biodegradable waste, a better way to dispose plastic will be its consumption by blending it in a material that itself has a very large volume. Since construction is the largest industry on the earth consuming billions of tons of natural and processed material, the use of plastic waste in different ways by the construction industry could be one of the effective solutions to the safe disposal of waste plastics. Researchers from the construction industry have started novel experiments to assess the feasibility of plastic waste utilization in construction activities.

2.1.2 MOLECULAR STRUCTURE

If monomer changes, the resulting plastic not only changes its form but also the chemical, physical, and mechanical properties. This is the reason for the versatility of plastic for large numbers of applications. Thermosetting plastic and thermoplastic are the main divisions of plastics based on the molecular bonding and arrangement

DOI: 10.1201/9781032621340-2

of resin structures in the matrix. The resin compounds arrange themselves during melting–cooling–hardening process.

2.1.3   THERMOSETTING PLASTICS

Plastic prepared by thermal curing (heating) of oil/crude sources through a cross-linking of monomers of resins resulting in a highly dense and strong three-dimensional molecular structure is referred as thermosetting plastic. Such plastics are incapable of getting remolded or reshaped due to the strong molecular bonding.

2.1.4   THERMOPLASTICS

Linear links formed by molecules of the compounds or resins that remain independent produce a soft and highly elastic plastic matrix during temperature curing referred to as thermoplastics. Chemically, the molecules are independent and maintain a linear or chain-like structure. The molecular structures are three-dimensional and cross-linked to each other, resulting in highly dense and weighted matrix. The molecules are cross-linked and can create many bonds through a single bond.

2.1.5   RESIN TYPES

A resin is a monomer used to produce links of polymer material. Following are the major types of plastics produced based on resin type:

a. PET plastics
   Polyethylene terephthalate (PET) is a thermoplastic primarily used for food-grade plastics. Therefore, the quality of the product should be maintained and not more than one-time recycling should be preferred. More repeated numbers of cycling alter the quality of the plastic. The primary resin used in PET is generally ester. Their primary properties are as follows:
   • High impact resistance
   • Exhibits crystalline structure and transparent solid form
   • Convenient to convert chemical properties
   • Suitable for recycling with or without chemical alterations
   • Feasible to render in flakes form for recycling

b. HDPE plastics
   High-density polyethylene is produced from ethylene monomer resin. Thermoplastics offer excellent strength and could resist high temperatures against melting. The following are the primary properties:
   • Highest strength to density ratio
   • Density up to 0.97 g/cm^3
   • High tensile strength
   • Convenient for recycling

c. PVC
   Polyvinyl chlorides are the second largest used polymers for plastic manufacturing. PVC may be produced rigid or flexible by nature. Most common applications are pipes, frames of doors and windows, wire and cables, and,

in some cases, bottles. PVC is classified based on its molecular weight as low, medium, and high. The properties are as follows:

- High density up to 1.45 g/cm³
- High hardness
- Does not offer good thermal and fire resistance
- Provides good insulation properties
- Medicine containers and pouches

d. LDPE

Low-density polyethylene polymers are the monomers of ethylene and are preferred for packaging. LDPE exhibits high flexibility in physical form and is non-reactive at room temperature. LDPEs are produced opaque, translucent, and with layering of additional materials, for example, aluminum is highly convenient at the stage of manufacturing of plastic films. Such aspects make them preferred material for packaging food stuff for a long retention period. Leading properties are:

- Low density
- Excellent chemical resistance for acids, alcohol, bases, etc.
- Capacity to offer versatility for films, sheets, and rigid containers
- Resistance to oxygen and moisture is excellent
- Good deformability and elasticity
- Corrosion resistant

e. Polypropylene (PP)

Polypropylene contains propylene as monomer and is widely used for packaging and labeling. The average global production of PP was reported as 55 million metric tons in the year 2013. Many of the properties are like polyethylene in suspension state. They contain the methyl group when compared with ethylene. This addition improves the mechanical and chemical properties of PP. Many times it is copolymerized with ethylene to achieve toughness and flexibility. At a very low molecular weight, they produce better plastic products. PP is likely to degrade by the effects of UV rays when in direct contact with the sun for a prolonged time.

PP is most conveniently used in film forms. Commonly BOPP films are used with additional coats of metalized or other plastic layers. Such films are widely used for packaging of snack foods. They maintain the freshness and quality of food by protecting the food from moisture. They are frequently used as an alternative to PVC for molding purposes. Following are some of the properties:

- Excellent flexibility
- Improved toughness
- Chemical resistant
- Non-reactive to bases and acids at room temperature
- Moldable
- Works good with other polymers
- Best resistance against fatigue
- Highly economical
- Compatible with ABS or others
- Could be colored with pigments and could be molded and starched

f. Polystyrene (PS)

Polystyrene is made by styrene monomers as foam or solids. Unlike other plastics, PS exhibits a clear and hard but brittle nature and is solid at room temperature but starts flowing beyond 100°C. The rate of biodegradation is low for PS products. Some of the properties are:

• Clear and brittle
• Suitable for foam state, which is an additional benefit over other types
• Exhibits average water and vapor resistance
• Styrene monomers change molecular structures when heated, therefore they exhibit excellent strength after extrusion process

g. Others

Modern plastics, namely ABS and elastomers, are included in this category. Such types include polycarbonates, acrylics, bioplastics, etc. When such plastics are used for making products, the recycling symbol indicates that they are still recyclable and environmentally friendly. However, the quality of such plastics depends on the manufacturing treatment and processes as well as the raw material quality.

2.1.6 Usage of Plastic

Globally produced plastics are utilized in a variety of activities. As discussed earlier, the flexibility of plastic as a material to be produced in any form has made plastic the most versatile material of the current times. Practically plastic contributes to the production of unlimited articles. The applications of plastics may vary nation wise and region wise, but plastic products have become common in nearly all parts of the world nowadays.

The maximum usage of plastics in packaging activities is up to 39.5% followed by building construction activities of about 20.1%. The electrical and electronics sector and the automobile sector are the third largest sections using plastic products with about 10% to 15% contribution. Information published yearly by the *Plastic Europe* organization may be referred to for detailed information on such statistical data. The report includes the extended usage of plastics in other activities like medical, furniture, and also agriculture.

Based on resin usage, plastics are utilized in a variety of activities and production of articles. The type of application or the production decides the use of the specific resin type. For example, water bottles are prepared from PET resin type of plastics. Of the overall resin usage, bottle production covers 44% share of the resin usage. Medical, automobile, household, and construction are the sectors that contribute to the use of high-density polyethylene (HDPE) and low-density polyethylene (LDPE) resins, with the average contribution of 10% of the total usage.

PE, PP, and PET are the primary types of resin utilized for numerous plastic products. They all are utilized at par in quantity; however, as discussed previously, about 50% of the total plastic production is used by packaging activities and thereby the usage of PP and PE has been predominant among the rest of the types of resins. In the following section, the application of plastics in construction-oriented activities has been discussed.

2.1.7   Building and Construction Activities

General

It is reported that about 23% of the total plastic production is utilized by building and construction activities and applications. This could be the second largest usage of plastic in an industry after packaging. As discussed earlier, owing to the large varieties of plastics by properties and by tier forms, their usage has shown to rise in numerous construction activities.

2.1.7.1   Applications of Plastics in Building Construction

- For moisture control in foundations and cellar walls
- For flooring and roofing
- As insulation barriers in majority of the components of buildings
- Pipes and conduits for fresh and waste waters, sanitation, wiring
- Coating for metal parts, sheets, rods, connections, wires, or cables
- Frames of nonstructural components, namely doors and windows
- Fittings and fasteners for non-load-bearing members
- Small elements like knobs, handles, switches, rollers, etc.
- For household furniture of all types
- For shuttering, formwork for reinforced concrete members
- As partial replacement of natural constituents in concrete like recycled aggregates, sand or fine particles, reinforcing fibers, and rebars
- For external strengthening of damaged elements by means of laminated sheets and as mixture of glass fibers and synthetic fibers

2.1.7.2   Types of Plastics for Building Construction Activities

The conventional and modern types of plastics are in parallel usage. They are in different forms, namely solid and solutions, according to the applications. Both the types of thermoplastic and thermosetting plastics are used in different applications for construction. The following are the plastic types bearing wide application scope:

- PVC: In sandwich panels, foams, sprays, adhesives, and tapes.
- Polyester: Sanitary wares, roofing and flooring, insulation, corrugated sheets, domes, etc.
- ABS: For minor fittings and panels.
- PE: Dampproof course and plumbing.
- PP: Water tanks and drainage pipes.
- Acrylic resins and nylon: Sanitary fittings and fixtures.
- Melamine: Panels and laminates.

Other than those listed above, plastics are used in varied forms directly or application based. Plastics are used in insulation, weather proofing and coatings, modern prefabricated panels, doors, and window frames as the most common applications. Plastic is being utilized extensively in the construction activities as material and also for making equipments. It is interesting to note the advantages and limitations of plastics used for construction applications.

2.1.7.3   Advantages and Convenience of Plastic
Usage in Building Construction

- Durability: Plastic exhibits good resistance against wear and tear, better moisture and corrosion resistance, and is less sensitive to impact and abrasion effects for a considerably long span. These attributes make it the most popular material for construction.
- Economy: Plastic offers good savings of materials compared to the conventional material usage. In manufacturing, low maintenance requirements and raw material cost and availability make plastics an economical option comparatively.
- Energy concerns and recycling: Many of the plastics used in construction are recyclable and reusable repetitively. The conventional materials are difficult to reuse because they are in complex form after demolition. To separate useful components from the combined mass of materials, multiple processes are required and therefore it is energy intensive.
- Flexibility of usage and applications: With the large variety of products plastics are highly suitable for varied applications. They could be produced with desired strength and shapes. Moreover, they are available in solid, liquid, and semi-solid forms. Plastics are produced in various forms such as sheets, laminates, sprays, foam, molded shapes, walls, rods, wires, and any similar other forms. Such a wide range of choices makes plastics a preferable material for construction usage.

2.1.7.4   Limitations of Plastic Usage for Construction Activities

Although plastics are highly versatile material for construction activities, they do have some limitations. Plastics are not fire resistant. Plastics are likely to get degraded and disintegrated under the constant exposure to the UV rays of the sun. Thermoplastics exhibit creep and loss of strength with time and altered temperatures. Owing to the high thermal expansion characteristics, plastic materials require proper design and allowance for expansion. Plastic could be one of the materials capable of providing acceptable strength, durability, and good flexibility for appropriate usage in many construction activities. With the innovations and research in the field of plastic industry, they could become even more effective and efficient material for construction activities in the future.

2.2   TRENDS OF WASTE GENERATION AND
THEIR HAZARDOUS IMPACTS

Plastic production is one of the fastest growing industries in the global market. In the last 60 years, plastic production has reached around 400 million tons per annum globally. Global plastic production has remarkably increased, with the increased production in Asian countries accounting for about 39% of the global production. EU shares about 25% of global production and collectively, while China alone produces about 15% of global plastics. It is followed by Germany at 8% from EU production. The production scenario is expected to increase within the next decade by 65% and it is expected that the production will reach around 1,600 million metric tons per year in the next three decades.

Production has been varying in the years from 2015 to 2022. In due course, the global plastic demand has remained constant with the increased awareness about the hazards and threats imposed on the environment and thereby mankind. The plastic demand has been constant in some parts of America. Minor increase and decrease are observed in the South Asian region and in Japan respectively. The production scenario is scattered as observed from the data; however, a constant growth in the world's plastic demand and production has been reported. The rate of increase of plastic production is observed at about 5%.

Plastic production scenario is influenced by the following primary attributes:

- Growth of population
- Changed lifestyles and modernization
- Industrialization and rural development
- Flexible chemical and mechanical properties of plastics
- Potential for global market and business opportunities

With the growth in the population and changed lifestyles of consumers, there has been a drastic change noticed in the types of plastics produced for new articles including packaging, construction, automobiles, and electrical and electronics pre-cuts. This has been discussed in the upcoming sections in detail.

2.2.1 PLASTIC WASTE

Plastic waste is separated into two major divisions based on the generation sources. The first type is called pre-consumer plastic waste. These are the plastic wastes produced in excess during the manufacturing of main articles. Many a times they are referred to as unfinished or unwanted portions of plastics produced in factories. However, the pre-consumer plastic wastes are not significant by proportion as it may turn into uneconomical production. The second and main type is called post-consumer plastic waste (PCPW). Post-consumer activities are primarily responsible for most of the portion of plastic waste generation. The PCPW is discussed in detail in this section.

2.2.2 PRIMARY REASONS OF PLASTIC WASTE GENERATION

The following are the primary reasons for waste generation by plastics:

- Plastics are non-biodegradable materials.
- Plastics exhibit excellent durability and a long life cycle.
- Plastics have a global production of about 400 million tons per year and only up to 30% are being recycled and reused. The rest is discarded as waste in land and water.
- Recycling and reuse of plastic waste are not yet effectively implemented globally.

- Global trading of waste plastics is less than 5% of the total annual production of plastic. This shows that very small amount of plastic waste is being utilized for reproduction of plastics and the waste proportion is still increasing.
- Up to 10% of total municipal solid waste is shared by plastic wastes. This proportion varies with the countries and their geographical locations. Owing to the highly mixed nature of plastic wastes, most of them are dumped in landfill areas.
- Lack of active consumer and user participation in waste plastic mitigation initiatives.
- More than 50% of total plastic production results in immediate waste after a single usage by consumers.

2.2.3 POST-CONSUMER PLASTIC WASTE (PCPW) RESOURCES

Plastic discarded after its intended purpose becomes plastic waste. Practically any activity utilizing plastic is a source of plastic waste irrespective of the scale and size of the activity. However, some of the primary sources of high-impact sources for plastic waste and types are:

- Municipal solid wastes
- Packaging
- Construction demolition wastes
- Scrape of automotives
- Agricultural activities
- Waste from industrial units and manufacturing
- Wastes from treated plastics in recycling or reuse
- Electronics and electrical appliances

2.2.4 PCPW GENERATION SCENARIO

Unlike production data on plastics, it is challenging to collect accurate waste generation data from plastic products. The reason behind this is the highly scattered waste generation trends and data collection methods.

The obvious plastic waste source is discarded packaging plastics. In EU countries alone, the MSW generated is 520 kg per person per year and it is projected to increase by up to 620 kg by 2025. About 40% of the total volume of MSW is shared by plastic wastes and therefore, increased MSW may also result in increased plastic waste. MSW waste plastic consists of about 70% wastes from packaging with a significant share of PE, PS, and PET resins including LDPE- and HDPE-based products. It is reported that shorter usage of plastic results in more waste.

2.2.5 HAZARDS OF PLASTIC WASTE

General

Plastic materials are produced for various usages and according to the ever-changing demands of industry and consumer needs. Primary types of plastics are thermosetting plastic and thermoplastics. A durable long life of plastic products has resulted in accumulation of huge plastic waste. Plastics are difficult for natural degradation and require special management and control for safe disposal.

Land hazards: Plastic releases toxic chemicals on contact with soil surfaces and subsoil strata. Chlorine is the main chemical compound harmful to the fertility of land. Landfill method of plastic disposal is a huge contributor to land pollution by plastic waste. It is observed that integrated plastic debris in micro and macro sizes exist with soil particles for a long duration. This presence of plastic contaminates the fertility of agricultural lands. In the landfill sites, plastics are slowly broken down by microorganisms of bacteria.

Plastic wastes of packaging and electronic and electrical applications are major contributors to land waste. Plastic littering is the primary reason for land plastic waste.

Water hazards: Oceans and natural water streams like rivers and ponds are constantly facing waste plastic debris accumulated over a long period of time. Waste due to unmanaged, littered, and non-recyclable refused plastics is the primary reason for water pollution. The plastics wastes from MSW are light in weight and could get transported to water streams and oceans easily and get accumulated on the water surfaces. Plastics are insoluble in water and do not degrade naturally. It is estimated that about 100 million tons of plastics are accumulated in oceans and around the coastal areas.

Air hazards: Of the various methods available for plastic waste management, direct incineration is the oldest method of disposal. With the advancement of technology in plastic waste posttreatment, direct incineration is comparatively less. This is due to the release of toxic gases into the environment and harmful side effects on the health of ecosystem in the surrounding areas.

2.2.6 HAZARDS TO LIVING ORGANISMS

Plastics release chemical compounds that are generally toxic in nature for the human body. Any source of plastic waste including land, water, and gases released in air affects skin, glands, and respiratory system. Ocean plastic waste debris accumulated on the surface of water can reach the sea bird's stomach indirectly when they consume other species that have eaten the plastics. Researchers have found heavy plastic waste from the dead birds' abdominal cavities. Plastic reduces digestive power, causes starvation and death. Littering of MSW plastic wastes is a major source of hazardous health effects of stray animals.

Overall, it is reported that plastic wastes in any form at any location creates serious hazards to the ecosystem unless handled, managed, and disposed meticulously. Toxic chemicals could create permanent hazards to not only the human body but also to other living organisms.

2.2.7 PLASTIC WASTE MANAGEMENT AND DISPOSAL

General

Plastic waste varies in types, forms, and sources of generation. Therefore, specific waste management and disposal methods are required. Globally, there are standard methods being practiced as mentioned below for plastic waste management and disposal, namely Recycling, Recovery, Landfill, and Reuse.

2.2.7.1   Recycling

Recycling of plastic waste is one of the most effective methods of waste plastic management. Recycling stands for the process of converting the waste materials into a new material in the desired form to be reused. In other words, it is a reproduction of the waste material with altered characteristics and properties. Plastic recycling is a process that includes collection, sorting, machining, and producing new material by utilizing plastic wastes. Recycling is processed based on the international plastic resin recycling codes ranging from 1 to 7. The most favorable plastic recycling code is 1–5 and for certain amount it is 6. The plastics prepared with code 7 are hard, rare, and difficult to recycle. They are sent to the dump yards largely. Recycling of plastic could be practically applied to most types of plastics like PET, PP, PE, LDPE, and HDPE. About 30% of total plastic waste is being recycled globally. The rate of recycling is increasing with modern methods and machines' availability. However, the following are some of the factors affecting the recycling of waste plastics in general:

- Quantum of plastic waste for the recycling facility
- Resin type and recycling code of waste plastics
- Sorting facilities and pre-treatments at the plant
- Source of waste plastic, namely industrial or municipal solid wastes
- Mechanical capacity of the plant
- Geographical location of facility and access to the facility

Recycled plastics have large market options. Recycled products are slowly being accepted and utilized for many applications, for example, making of new plastic products partially; the recycled plastic grains are being mixed into molten plastic and resins. The recycled plastic aggregates have been used in secondary products or less important products such as carry bags and frames or handles of household cutlery.

Even with the modern techniques and setup available for recycling units, the effectiveness of recycling is not attained in many parts of the world. The limiting factors include awareness of users on plastic disposal, suitability of waste plastic for recycling, and market for recycled products. Specifically, packaging wastes are not fit for recycling and require alternatives for disposal. The current practice for packaging waste disposal is landfill. Dumping of flexible waste plastics continues to raise pollution after disposal for prolonged duration, as the microbial activities release gases and the small fraction of plastic remains present in the waste streams.

2.2.7.2   Recovery

General

Waste is commonly disposed in landfill as its ultimate destination. The modern concept of utilization of waste in producing useful heat, electricity, or as an alternative to conventional fuel is termed energy recovery. With the development and improvement in environmental protection policies and laws, energy recovery has been emphasized globally. It is also known as non-hazardous waste management. Plastic waste is used as alternate fuel in industrial processes and reused for making recycled plastics following the concept of energy recovery. This method of plastic waste management has been proved economical, less hazardous, and acceptable by many countries in the world.

2.2.7.3 Advantages of Energy Recovery by Waste Plastics

Several benefits are observed by employing the energy recovery option with waste plastic as a resource material in different sectors such as partial replacement of conventional fuel, reduction in landfill dumping, and better fuel option for direct and feedstock incineration. There are two methods of availing the benefits of energy recovery out of waste plastic sources: (a) by direct incineration of waste plastic and (b) partial replacement of conventional fuel like coal and oil.

2.2.7.3.1 Landfill

The landfill includes waste that are primarily wastes due to littering and those that are difficult to collect in a systematic manner. It contains both dry and wet types of waste together. After a preliminary sorting process, the waste is diverted to the landfill. When the predicted height of dump is obtained, the site for landfill is relocated to a new destination. In the line of such facts landfill is being preferred as the last option for waste disposal in case of non-availability of any better and effective option. With the context of plastic wastes landfill is not an environmentally friendly method of disposal. The following are the drawbacks of landfill disposal of plastic wastes:

- Plastic wastes are dumped without sufficient pretreatment, and they are primarily non-biodegradable in nature, creating water and land pollution.
- Plastic waste dumped in the landfill remains in the top and sub soils for a considerably long duration, this is the reason for water clogging, reduced soil fertility, and altered mineral contents of the agricultural lands in surrounding areas. Many plastics degrade in the presence of UV rays of the sun. However micro-sized plastic particles remain in the inner layers of soil and pollute it.
- Landfill areas are open to the sky and could give free access to lightweight plastics, which fly away and pollute the nearby ecosystem of land and water streams.
- Plastic waste from the landfill areas gets mixed into the nearby water sources and increases pollution and water borne diseases.

2.2.7.3.2 Reuse

Utilization of plastic waste into an activity or process either directly or as an indirect application that could reduce the accumulation of waste is referred as reuse. Reuse is a broad term describing the methods or treatments for effective waste utilization by industry and consumers, which could reduce the generation of plastic waste. The reuse of waste plastics includes recycling of PET bottles, use of sorted and recyclable plastic wastes, preventing the refusal of plastics, and encouraging repeated use, partially or completely replacing the conventional fuel by plastic waste for generating heat and temperature.

2.3 FEASIBILITY OF PLASTIC WASTE TYPES FOR ADDITION IN CONCRETE COMPOSITE

General

Effective management of plastic waste requires innovative methods. Initiatives by industries and public sectors like municipal corporations are still in the developing stages. Policies on plastic waste disposal have been effectively refined and strictly implemented in many countries. But the success ratio has not been more than 60%. Efforts have been carried out by construction industries to mitigate hazardous impacts of plastic wastes by innovatively utilizing them as partial replacement of conventional materials and fuels. Energy recovery by using waste plastics with partial replacement of conventional fuel like coal has been successful up to an extent. Concrete is a material being consumed globally on a large scale with 400 million tons annually. Researchers have investigated possible usage of waste plastics in concrete to avail a dual benefit of development of sustainable construction material and waste utilization. Listed below are some of the examples of utilization of waste plastics in concrete:

- As constituents in cement concrete for preparing green concretes
- As constituents in rigid pavements
- Fuel for cement manufacturing

2.3.1 GREEN CONCRETE

Green concrete is a modern term used for concrete prepared by waste materials as partial replacement of conventional constituents. The way of waste utilization in concrete extends the following benefits:

- Reduced demand for conventional materials
- Reduced energy demand required during manufacturing
- Safe disposal of wastes causing hazards to the environment
- Mitigation of littering of waste plastics
- Reduced waste plastic dump into the landfill
- Economical solution of waste utilization for the specific wastes not suitable for recycling, namely metalized plastic wastes from food packaging
- Changed properties could improve some of the conventional response of concrete for strength and durability

Compared to normal concrete, green concrete tends to be more environmentally friendly, while providing parallel benefits of savings of conventional materials and in some cases better results for properties.

2.3.2 SCOPE OF PREPARING GREEN CONCRETE BY POST-CONSUMER PLASTIC WASTES

Micro-sized plastic fibers from virgin raw materials have been mixed in concrete to enhance crack resistance and durability aspects in fiber-reinforced concretes.

Waste plastics are also recycled into macro-sized fibers and flake form and mixed with concrete as constituents by fractions of total concrete volume. Researchers have reported significant improvement in splitting tensile resistance and resistance to shrinkage cracks and creep behavior due to the addition of fibers retrieved from the waste plastics. These novel experiments have shown potential to the concept of effective utilization of waste plastic as concrete constituent. Even though the dosages are small, they have exhibited significant effects on concrete properties. Recycling of waste plastics could produce granular form as well. Such granules are used as the source of aggregates of varying size and partially replaced with natural aggregates in concrete mixes. Extensive research work could be carried out for acquiring the importance and success of another way of utilizing the waste plastic in concrete.

2.3.3   RIGID PAVEMENTS

Rigid pavements are widely used for roads and runways. The top surfaces or the bearing courses are prepared by bitumen or cement concrete. Addition of waste plastic in flakes or fiber form has been initiated and is being practiced successfully. Inclusion of recycled waste plastic fibers contributes to reducing the shrinkage cracks and surface cracks in pavements. Pavements are under constant exposure to temperature variation, resulting in surface and micro-cracks in the hardened layer. The presence of plastic contributes to protecting the crack propagation and prevents undesirable deformation of the surface. Further, the pavements of runways are under constant effects of impact loading. Experimental study of impact resistance of concrete containing plastic waste fibers has shown improved impact resistance. Plastic fiber helps to improve the stress distribution in the hardened mass and reduces the stress concentration. Waste carry bags can be included in the hot bituminous concrete mix without recycling, exhibiting one of the simplest ways of utilizing plastic wastes.

2.3.4   FUEL FOR CEMENT MANUFACTURING

Cement manufacturing is an energy-intensive process. It is the mass production of material and requires a high amount of conventional fuel to attain the desired temperature to fuse the raw materials into clinkers. Replacing conventional fuel like coal with waste plastics recovered from the landfill could reduce the partial need of coal consumption. However, it could be difficult to assess the changes in the toxic gases released by the burning of such waste plastics.

2.3.5   POST-CONSUMER METALIZED PLASTIC WASTE (PCMPW)

General

Plastics layered with metallization called metalized plastic is a primary source of material for food packaging. The plastic waste generated by metalized food pouches can be termed as post-consumer metalized plastic waste (PCMPW). In this section, environmental concerns and hazards by PCMPW are discussed. Reused wrappers, sachets, pouches, packs, and flexible containers used in packaging solid, liquid, or food suspensions are the direct and primary sources of PCMPW. It is reported that about 50% of total plastic used in packaging is used in food packaging and results in immediate waste on a single usage. Food waste from houses, institutions, public places contain about 10% of PCMPW of solid waste.

2.3.5.1   Environmental Concerns by PCMPW

Following facts of PCMPW make them hazardous to the environment:

- Improved durability by metallization causes difficulty in natural degradation of the waste articles.
- Lighter weight makes PCMPW float on water surfaces, which eventually results in clogging of natural water drainage.
- PCMPW are highly non-reactive to chemical reactions. This increases the prolonged presence of particles of PCMPW in land, water, and air mediums.
- PCMPW gets accumulated and spreads over the landscapes, adversely affecting the fertility of soils.
- As PCMPW are not recyclable, repeated usage of base material is not possible. Therefore, constant usage of new raw material is required to produce metalized plastics. This adds to the emission of greenhouse gases and more usage of raw material.

2.3.5.2   Scope of Sustainable Usage of PCMPW

PCMPW is found littered in natural streams and pollutes valued natural ecosystem in many ways as discussed earlier. Instead of landfill or recovery by PCMPW, innovative methods can be developed for sustainable usage of PCMPW. In fact, the packaging industry extensively uses metalized films for many food articles and has been responsible for a constant growth in plastic waste that are unfit for recycling and reuse. It could be worth experimenting to utilize the PCMPW since the material is being extensively used in many countries. In summary, for more than 60 years plastic production and waste resulting from the use of plastic has been continuously increasing globally. Being the most innovative material of modern times, plastic has replaced many conventional materials, namely stone, metal, glass, and timber, and has become a part of our daily life. Plastic, on the other hand, possesses excellent durability properties and is difficult to degrade naturally. A constant demand in plastic and limited practice of safe disposal of the waste generated by plastic usage have raised an alarming environmental concern. Though there are many innovative methods and legislative steps have been implemented to control the situation, specific types of plastic waste such as PCMPW are still not sustainably disposed or utilized. Some of the major reasons for this are lack of consumer awareness, material characteristics, and absence of effective modern disposal techniques. However, research activities could be able to answer the question of safe disposal of such hazardous plastic wastes in future. Noticeably, researchers from the construction industry have already started efforts in this direction, as could be revealed from the literature available as discussed in the next chapter.

3 Industrial Wastes

3.1 TYPES OF INDUSTRIAL WASTES

General

The manufacturing process of any product is bound to generate various unwanted substances, sometimes referred to as by-product or simply waste. However, the by-products may or may not be utilized for other products, and the wastes are largely subjected to disposal. Depending upon the product and processing, the waste is generated and they vary in type, nature, properties, and form. One of the most significant parameters of the waste being generated in any industry is the quantity of waste. If a waste product is generated in a substantially large quantum, the disposal becomes a challenge. Especially in the absence of an effective disposal option, such wastes can create environmental concerns. In this section, some of the most significantly large industries and their waste products have been discussed and their feasibility for utilization in concrete has been talked over.

3.1.1 Types of Industries Producing Significant Quantity of Wastes

Generally, the industry consists of several processes, needs huge quantities of raw materials, and is involved in the mass production of the articles to produce quantified wastes, namely

- Thermal power stations
- Steel manufacturing plants
- Oil refineries
- Chemical manufacturing plants
- Mining activities
- Plastics and rubber industries
- Textile industries
- Wastewater treatment plants
- Electronics and electrical goods manufacturing
- Paper mills
- Agriculture and milk product-based industry

All the industries mentioned above and beyond the list produce at least three forms of waste, namely solid, liquid, and gases. The waste from any of these forms is produced when they are difficult to dispose of directly or consists of hazardous elements in their original forms. On the separation of the hazardous elements from the waste, the actual waste generated is to be disposed of alternatively but safely. From this point pollution caused by waste becomes a concern. The waste is often difficult to get disposed of in an environmentally friendly manner. Therefore, alternate uses of the

DOI: 10.1201/9781032621340-3

TABLE 3.1

Mapping of Wastes with Concrete Ingredients for Utilization

Construction Activity	Comparing Point	Waste Ingredient	Mapping Remarks
Concrete mortars	Strength, workability	Powder, granules	Chemical reaction
Bricks	Water absorption, strength, surface	Powder, granules	Water resistance
Structural concrete	Strength, durability, and workability	Powder form, dust form, granular form, liquid form	Chemical reaction, binding capacity
Pre-cast members	Strength, durability	Powder, granular	Stability under loading conditions
Roofing materials	Water, thermal, and impact resistance	Powder, dust, granules	Stability, durability
Road pavements	Impact, crack, freeze, thaw, melting resistance	Powder, dust, granules, pebbles, boulders	Stability, durability
Soil improvement	Bonding, stability, strength	Powder, granules, pebbles, boulders	Stability, durability
Slope stabilization	Bonding, stability, strength	Pebbles, boulders	Stability, density

waste products are required. Possibly, all wastes are useful for alternate utilization except the radioactive and nuclear reaction-related wastes; however, taking the scope of the book into consideration, this section restricts the discussion to industries whose wastes and by-products are useful for building and construction-oriented activities.

The construction and building industry is one of the capable sectors wherein wastes products may be utilized in a significantly large quantity. The construction of infrastructural members is made with composites and steel primarily. In addition, they also include the use of secondary materials, namely glass, plastic, and some new nonferrous materials. Therefore, aligning the requirements of the construction and industry wastes with needed strategies is a better way of assessing the feasibility. Table 3.1 illustrates how to map the construction activity, waste ingredient, and their common points of comparison for utilization. While mapping the waste against the ingredients, care should be taken about the factors influencing the selection of waste and also the material to be altered. Before mapping the attributes, check with the following influencing factors. The material to be altered will be referred to as primary material here.

3.1.2 Factors Influencing the Waste and Construction Mapping Materials

- Chemical composition of waste and primary material
- Intensity of the specific chemical compound in the waste and primary material

- Physical properties of the waste and primary material
- Physical form of the waste and primary material
- Confirmation of the presence of hazardous substance in the waste and primary material
- Response of the waste material and primary material to reaction with water
- Reaction of the waste and primary material to thermal sensitivity
- Response of the waste and primary material to the general human response systems, namely direct and indirect contacts of the body

Above are a few of the influential parameters to be considered while selecting the wastes and mapping them with the primary construction material. Table 3.1 illustrates the process of mapping the materials.

Kindly note that the above table can still expand with more options and possibilities of applications, comparisons, and utilization as the field and area are highly expandable and consist of several possibilities. The only concern should be the waste type, and the material being replaced should be compared in as much detail as possible for exact utilization and advantage expected from the modification.

3.1.3 Selected Industrial Wastes and Their Utilization: Examples

Discussion of the technical and scientific facts to utilize industrial waste may not be completely clear. Let us explore some examples.

a. Fly ash and bottom ash: This industry's waste is generated in a huge and constant quantity because it is produced in thermal power stations and these units never stop running; therefore, the burning of fuels is a continuous process.

The following are the primary reasons to use fly ash in construction activities:

Fly ash and bottom ash consist of fine particulate matter with good possibility to be added into the conventional material such as cement in powder form.

The finer particle size of the ash provides a better filled matrix in the concrete and thereby improves the intermolecular structure of the mix.

The chemical analysis of the ash samples of both fly ash and bottom ash shows that they are mineral-rich materials. Fly ash consists of calcium and silicon ingredients with maximum contribution; therefore, they can become an integral part of the cement, cement-based mortar, and cement-based concrete.

Being the source of silica-rich and calcium-rich oxides, the utilization of ash has become possible in the development of green or sustainable products, namely alkali-activated binders and matrix prepared with natural or recycled aggregates thereof.

The ash may be processed further to obtain nano-sized particles for the development of efficient and versatile mixes.

Fly ash and bottom ash are good ingredients for roads and highway layers. They provide better surface, anchorage, and integration with the other bituminous concrete materials. The adhesion capacity increases with the presence of fly ash for binder materials.

Fly ash is an excellent powder form of pozzolanic materials, capable of providing resistance to the water ingress and other fluid entry into the surfaces. Therefore, they are suitable for plastering and lining canals and tunnels.

The approximate average annual global production of fly ash and bottom ash is 500 million tons, which is more than any other construction material. Such a huge quantity indicates that the material can be obtained in sufficient quantity for any amount of the replacement and for extended periods also.

Fly ash and bottom ash are highly adaptive in nature and easily get mixed with other constituents and minerals, namely clay, cement, soil, liquids, and viscous fluids. This ability opens several opportunities for the development of sustainable construction materials.

Fly ash can be utilized as the chemical compound for chemical processes namely pyrolysis and can function as an antitoxic substance in the sludge stabilization for the treatment of wastewater.

Above are a few of the reasons to utilize fly and bottom ash in the construction industry.

The general chemical constitution of type F fly ash with different sources has been identified by employing microstructural chemical analysis methods. Fly ash consists of 60% of silica and is a rich source of silica-based pozzolanic material. There are other chemical constituents also available, namely ferrous, alumina, and calcium in oxide forms, but all these are in negligible quantities. However, it is to be noted that the chemical concentration will depend on the type of fuel sources being used in the industrial processes, such as bituminous or lignite-based fuel. The fuel type will affect the concentration of the materials significantly. For example, fly ash produced by lignite-based sources will consist of calcium oxide of about 40%; however, the bituminous fuel will consist of only 10% of the calcium oxide content. The composition indicates that depending on the source of the fuel and parent rock source, the burning results in different constituents in varying proportions.

The concentration of silica and calcium varies significantly in a given ash sample since it depends on the type of the fuel used in the power generation process. Therefore, before the addition or replacement of the conventional material, it is important to carry out chemical analysis of the fly ash or bottom ash sample. General practice suggests that fly ash consisting of rich silica components is more useful in the concrete making materials, namely cement mortars, wherein the role of the fly ash is to provide more silicates to the bond generated by the cement and water and resulting in better intermolecular bonding of the constituents.

3.1.4 EXPERIMENTAL EVALUATION AND PERFORMANCE

Several technical and scientific assessments and evaluations of concrete consisting of fly ash are available to refer and understand the working and contribution of fly ash in concrete-based material development. The presence of fly ash in concrete has contributed to improved mechanical strength, permeability resistance, reduced chemical ingress, enhanced intermolecular structure of the mortar, and effective resistance to severe environmental conditions.

The fly ash particles are regular and consistent in shape and provide larger surface area. Fly ash can be found to have particle sizes in the range of 10–100 μm, while cement particles are in varying forms and shapes and exhibit the same size range as in case of fly ash, therefore the replacement of the cement by fly ash becomes effective and does not alter the matrix. However, it is to be noted that fly ash does not carry a binding capacity and therefore does not participate in the active chemical process with water.

3.1.5 PLASTIC AND RUBBER INDUSTRIES

Plastic as a material and product has become one of the integrated parts of daily life globally. Now, plastic is nearly difficult to manage when it turns into a waste. There are several attributes of waste plastics, and globally the policies, rules, and processes have evolved to manage and dispose of plastic waste. However, the safe disposal of plastic waste has remained a challenge worldwide.

The fact sheet of plastic production and usage suggests that not all plastic wastes are disposed safely. This includes reuse and recycling also. Therefore, researchers have made efforts to include the selected plastic and rubber wastes in construction practices. The following are some examples.

3.1.6 USE OF WASTE PLASTIC IN ROAD CONSTRUCTION

Several research papers deal with used plastic bags in pavements. For the Indian context, the railway ministry has published a guide for use of plastic in road construction and the suggested mix design requirements have been proposed by the government. The guideline consists of the types of plastic waste containing nearly all types of resins and the form of the waste plastic type. This is because the segregation of the plastic waste is a difficult task and there are a few methods available to separate the mixed wastes.

Waste plastics are now being used in road making materials. The flakes have been replaced with fine and coarse aggregates of conventional bitumen mortar used in flexible pavement. The test results suggested that up to 2 tons of waste plastic can be utilized for 1 km length of about 5 cm thick road covered with flexible pavement. There are several references available indicating the utilization of plastic in construction applications. It is to be noted that the mixed nature of plastic types may present a challenge, therefore the recycled waste should be selected carefully and must be available in varying sizes for appropriate replacement of conventional ingredients.

3.1.7   Use of Waste Rubber in Concrete and Composites

Like plastics, rubber wastes have also been explored for their possible inclusion in concrete and similar composites. One of the major sources of waste rubber is the waste tires of vehicles. The waste tires pose several environmental challenges as they are difficult to completely recycle. Researchers have tried to use the rubber and wires from the used tires in many forms. Coarse and fine aggregates have been used to replace conventional aggregates, moreover crumb rubber in the form of macro sized fiber like form has been used as a filler material and to an extent as the fiber in the concrete mixture. It is possible to convert the waste tires into different forms and sizes of the particles and products. These forms have been extensively utilized in the making of construction composites. One of the common forms of waste rubber used is the granules and crumb rubber. The experimental results have established that the rubber waste may be blended along with the other wastes, namely demolition wastes to obtain increased durability in the concrete.

3.1.8   Wastewater Treatment Plants

Though this segment may not be called an actual industry, the production of waste is significant in quantity and also these facilities are generally found in most parts of the world. The waste products from the water treatment plants are treated water and sludge. Both solid and liquid wastes are finally disposed into water streams after ensuring the removal of the polluting components as per the allowable limit. However, as water is one of the most important resources, there should be an effort to conserve it. Researchers have made efforts for the possible utilization of different wastes in concrete and cement-based composites in the past years. The wastes have been utilized to produce cement and soil-mixed bricks, pre-cast elements, and other similar applications. The sludge from the plants have been dried and crushed into varying sizes to obtain fine aggregates for mixing in concrete. Efforts have also been made to retrieve the fine aggregates and replace them with conventional concrete making materials up to at least 5% of the total material as sand. However, these products require more detailed study for long-term durability as there is always the presence of chemically active constituents.

In summary, industrial wastes have been explored to an extent for their possible inclusion in the concrete and construction composites through research and development.

3.2   A SCENARIO OF GENERATION AND UTILIZATION

General

There are several large-scale industries generating huge number of wastes, and statistical data on the topic is noteworthy. The present section deals with the fact sheets regarding the water generated in different avenues. Alongside a utilization scenario, the related challenges and requirements are also discussed in this section. The details are shared from the sources available to access at the national and to some international levels primarily.

3.2.1 Waste Generation from Thermal Power Stations

3.2.1.1 Generation

The primary waste generated from the power stations is of several types including recyclable and non-recyclable waste. The waste is found in solid and liquid forms or as mixed type. Few of the hazardous wastes including chemicals and oils are generally disposed of by commercial methods. All thermal power plants have dedicated treatment and storage disposal facilities available wherein the non-recyclable wastes are finally disposed. The waste materials that may be available for construction utilization are largely the ash developed during the burning of fuel in the form of bottom or fly ash as discussed earlier. It may be observed that the most obtainable waste is ash and water. The wastewater from the plant may be safely disposed of to nearby natural streams after filtration treatment; however, the ash requires effective reuse. Nearly 80% of the coal fuel is converted into ash during power generation. Approximately 122 million tons of fly ash are produced annually, where large-scale waste is produced and requires to be reused.

3.2.1.2 Utilization

The references suggest that fly ash is used in cement manufacturing, concrete making, alternate construction composite making, pre-cast members, soil stabilization, and in the making of mineral admixtures. Varying proportions of fly ash of between 10% and 30% by weight have now become a commonly available binder in cement production. The cement prepared with fly ash and other similar pozzolanic materials is termed as Pozzolanic Portland cement (PPC). Modified cement has become commercially available and is being utilized extensively. Though there are a few concerns on its usage, PPC is steadily replacing OPC in several applications.

3.2.2 Steel Manufacturing Plants and Wastes

3.2.2.1 Generation

Steel is extensively being used in the construction and manufacturing industries. Though the material itself is recyclable and shows up to 97% efficiency for reuse in most of the products, the processing of steel is energy intensive. A steel plant produces slag, contaminated sludge, mill scale, refractories, and scrap in relatively large scale. Though the waste is largely utilized in many applications, however, sustainable usage is to be focused where the utilization of the steel waste does not demand energy with higher intensity. Overall, in a steel plant, 1 ton of steel requires raw material of about 2.8 tons, nearly 2.5 tons of water, and air of 5 tons. The manufacturing process can produce up to 8 tons of wet dust-laden gases, half a ton of contaminated water, and up to 0.8 tons of solid waste.

3.2.2.2 Utilization

Slag is one of the most used wastes for alternative use. Slag is commonly used in manufacturing construction materials and building-related items. With the recent development of alkali-activated concrete, the binder components namely cementitious powder are being developed using fly ash and granulated grounded blast furnace slag. The binder so produced is generally referred to as geopolymer binders. It is being claimed that it can replace the conventional ordinary Portland cement in concrete. Extensive studies have been carried out worldwide and the binder has shown promising response. Other than this, the slag is also utilized in concrete as aggregates of varying sizes, namely coarse and fine particles. The feasibility of such replacement has been explored by researchers and found suitable for its intended purpose.

3.3 FEASIBILITY OF INDUSTRIAL WASTE UTILIZATION IN CONCRETE COMPOSITES

Concrete composites have been extensively explored for the possible addition of different industrial wastes. Fortunately, the industrial wastes or its by-products may be correlated with the conventional materials to an extent. As mentioned in the earlier subsections of this chapter, there are many industrial waste materials that may qualify as cement concrete constituents. However, it is important to ensure that the addition of such materials fulfills the other essential requirements of the materials' performance and properties. The following is the suggested checklist useful to determine the adequacy of the waste materials in concrete as one of the constituents.

3.3.1 RESOURCE CONSISTENCY AND RELIABILITY OF THE WASTE MATERIALS

One of the major challenges in waste utilization in concrete is resource stability. In this regard, if the waste is from an industrial process, then it may be possible to have a consistent supply of the waste with the desired quantity. The constant supply may vary according to the type of waste. For example, class F fly ash is produced in thermal power stations wherein coal is used as the primary fuel. Similarly in the case of used foundry sand, they are of varying types namely green and bonded one. Any change in the source of such material will influence the characteristics of the composite being explored to a significant extent. The second aspect is the reliability of the waste in terms of purity and standards. Nowadays, the ashes are being pretreated before being utilized. This will benefit the waste to remain as pure as possible. The material impurities may not be easily removable by the laboratory work. Hence, one should check that the waste materials do not have any chemical or physical impurities.

3.3.2 CHEMICAL AND PHYSICAL PROPERTIES OF THE WASTE MATERIALS

The concrete composites are sensitive to the chemical processes, namely hydration of cement. If the cement fails to meet the standard requirements, it is difficult to obtain the desired result from the matrix. Similarly, when a waste is mixed into the matrix and especially as a replacement of conventional materials, their chemical composition must be examined. Many times, the waste consists of a few compositions in a significant proportion, namely class F fly ash is rich in silicon oxides while class C fly ash consists of calcium as a dominating component. There are always specific roles of each such component in the concrete making process.

The concrete prepared with class F fly ash provides excellent hydration and strength properties to the concrete, while class C fly ash results in excessive heat of hydration and may damage the intermolecular bonding of the hydrated paste and adhesion on the aggregates. The physical properties of waste material are also important. Namely the waste plastic fibers should be checked for their surface smoothness and the inherent tensile strength capacities before use in the concrete. This is important because the fiber-reinforced concrete may require excellent tensile resistance and cracking resistance in the hardened mass. Hence if the fibers are smooth and do not get anchorage in concrete, the desired strength against splitting actions may be difficult to achieve.

3.3.3 Suitability of Waste Materials for Pre- and Post-processing of Concrete Making

We all are aware that the concrete making process requires primary actions, namely dry and wet mixing, amalgamation, pouring and placing, vibrating, and curing in water. The addition of waste material should not interrupt these processes. Consider an example of plastic fibers again. They are light in weight and non-reactive to the chemicals. Due to the ultra-thin and lightweight properties, they should be added in the dry mix with care. The authors' experience of working with these fibers suggests that the mixing drum should be closed from the top during the rotating movement since the fibers may fly away in air as they are very lightweight. Similarly, when the water is added into the dry mix of concrete, the plastic waste fibers do not easily get mixed with the other constituents. Moreover, upon vibration of the specimens the fibers show a tendency to float around the top layers. A similar challenge is faced while working with very fine powder of marble waste, ashes obtained from the stubble residues, and other waste materials. Therefore, it is advisable to study the physical behavior and interaction mechanism of a particular waste with the concrete making processes at each stage.

3.3.4 Environmental Sensitivity and Typical Behavior of Waste in Specific Conditions

Concrete composites may be considered as durable material to an extent. Concrete deteriorates relatively faster than the other materials. However, the rate of deterioration varies with the quality and strength parameters of the hardened concrete; the structures built with concrete are bound to have cracks and permeability. The wastes namely fly ash, silica fume, and other industrial products may reduce the air voids in the hardened mass as also observed in the microstructural investigations of the concrete consisting of such wastes. The waste should be resistive to moisture and water absorption to an extent. Also, if the constituents like sand and aggregates are being replaced, the size and shape of the particles must be extensively studied before the application. The physical properties of waste materials may create significant effects on short- and long-term durability response of the structures. Therefore, the feasibility of the addition of the waste materials should be evaluated by experimental assessment of the specimens and by employing all possible replacement variables.

In summary, the readers should confirm the preliminary and primary properties of all the waste materials with the constituents under consideration for replacement. One more important aspect regarding the replacement ratio is the volume and weight. We all may have come across literature mentioning the replacement of conventional materials by waste material either by volume or by weight. There may be a question which should be practiced. The solution is in the physical form of the conventional material and the waste material. A simple way to address this issue is to focus on the density of the materials. Usually, cement, sand, and aggregates are replaced by their partial fraction of weight in percentage values with appropriate waste materials. For example, fly ash can be added into the concrete by the percentage weight of the cement. Similarly, the artificial sand may be replaced by weight percentage and so as in the case of aggregates. However, the challenge is with lightweight materials and plastics.

The lightweight materials namely expanded clay aggregates, foam granules, natural or cellulose-based agricultural wastes, and ash obtained from the incinerator thereof are a few examples of lightweight additives for concrete. In such cases, the addition may be employed by the volume fractions of the concrete mix or in a total quantity. In this book, you will come across the use of metalized waste plastic fibers, possessing relatively low density and therefore they have been added by the percentage of the total volume of the concrete mixture. This is necessary because the weights of conventional and lightweight waste materials are different and it may happen that in matching the weights, the volume of the lightweight waste materials gets increased to such an extent where the addition becomes practically unfit for the purpose.

Overall, waste materials are used for mixing, replacing, and addition provided they fulfill all the requirements and feasibility checks mentioned above. The next chapters focus on the utilization of the waste plastic and waste industrial material, fly ash and slag, and silica fume powders. The reader will notice that all the mixes are added with the wastes according to their own chemical, physical, and mechanical characteristics. This book focuses on the demonstration of how to include plastic wastes and other industrial wastes in a blending mode in the conventional concrete mixture. The readers are advised to carry out sufficient pilot studies to understand the behavior and responses by the wastes and the property alteration of the conventional concrete.

4 Development of Concrete Composite Using Plastic and Industrial Wastes

4.1 PROPERTIES OF WASTE PLASTICS USEFUL FOR CONCRETE MANUFACTURING

Waste plastics may be obtained in several forms, namely bags, wraps, small to large articles, packaging pouches, and so on and so forth. However, there should be a scientific way and method of using waste plastics in construction composite. In this book, two types of composites have been discussed. The first type belongs to the conventional concrete prepared with cement, aggregates, sand, and water as well as the admixtures, wherein the plastic wastes have been added as different forms, namely fibers, granules, powder, and aggregates of varying particle sizes. The second type is the composite developed without use of cement termed as "alkali-activated concrete" or commonly "geopolymer-based concrete." The latter type includes the industrial wastes as the main ingredients for making the binder material with the help of alkali activators as chemicals, namely sodium silicate and sodium hydroxide in liquid form. This section deals with the conventional concrete consisting of different types of plastic wastes in varying form, where the primary pre-treatments, properties, and similar information on usage of plastic wastes in concrete have been discussed.

The following are the plastic wastes utilized for performing the pilot studies. The objective of the pilot study was to identify the feasibility of the addition of the specific type of plastic in concrete.

4.1.1 METALLIZED PLASTIC WASTE

The metallized plastic was selected for addition in concrete as fibers. Plastics are one of the most challenging wastes causing several environmental concerns, because of the post-consumer activities. The rejected waste material of a specific plastic is always found mixed with other municipal solid wastes. Therefore, segregation of the plastic from the solid wastes is important.

Metallized plastic film produced by a packaging unit from the local industrial area was shredded into flakes of different sizes as shown in Figure 4.1. Tests for obtaining tensile strength of film were conducted on the film sample as shown in Figure 4.2. Typical properties of metallized polypropylene are listed in Table 4.1. Metallized plastic films were shredded into three sizes: type A of average 1 mm, type

DOI: 10.1201/9781032621340-4

FIGURE 4.1 Plastic waste shredding.

FIGURE 4.2 Thickness of metallized plastic film.

B of average 5 mm, and type C with 1 mm × 20 mm (W*L) dimensions. Type C size was regarded as fibers and the films were manually rendered into fibers. All types of fibers rendered from metallized thin plastic film are shown in Figure 4.3.

TABLE 4.1
General Properties of PCMPW

Sr. No.	Property/Description	Unit
1	Resin type	Polypropylene
2	Name	Metallized polypropylene
3	Recycling code	5
4	Density	0.925 kg/cm³
5	Measured tensile strength	1,000 N/mm²
6	Elongation	70%–80%
7	Thickness	0.08 mm

FIGURE 4.3 Shredded fibers.

The figures illustrate the types of fibers obtained from the metallized plastic wastes and how the primary tests have been conducted. Primarily, the thickness of the fibers and the average dimensions are important parameters. The thickness of fibers should be checked because the fibers consist of a larger aspect ratio as they are macro fibers. It is to be noted that the fibers are of two types: micro and macro scale for concrete.

4.1.2   Recycled Plastic Waste

Plastic waste that might have reached its end-of-life phase or has no potential application was collected from local PVC and HDP pipes' industries situated near Rajkot city. Waste has been collected in three different forms, which are represented in Figures 4.4–4.7.

Figures 4.4–4.7 show recycled plastics in different forms, namely powder, flakes, or coarse-sized aggregates, fine-sized granules, and rounded wires as fibers. The interesting part regarding above-mentioned plastics is that the waste of such plastics is easy to obtain and largely they are not found to be mixed with the other sources of

FIGURE 4.4   Crushed PVC pipes into powder form.

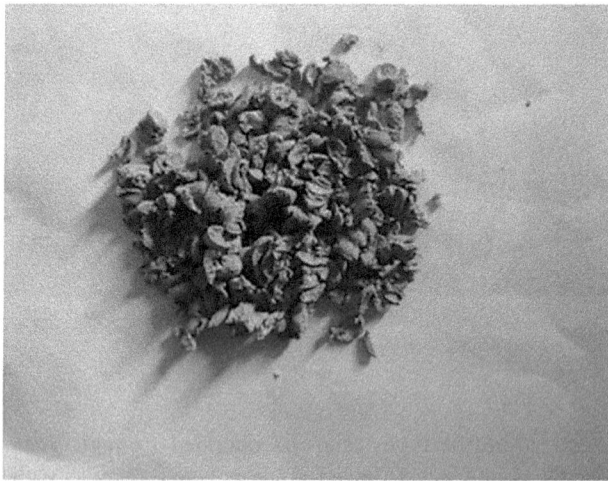

FIGURE 4.5   Recycled PVC flakes/aggregates.

waste. From building demolition sites, agricultural fields, and commercial units, such plastics are very convenient to be obtained. The waste is usually obtained in bulk from the waste collectors and sent to the recycling facilities. The recycling process includes initial separation of the useful articles, and a small amount of the plastic waste is rejected from the recycling process. The recyclers are mechanical machines primarily crushing the articles into smaller parts, washing them, melting them, and finally breaking them into large particles, small particles, or wires. The current study deals with the pilot study on three different forms in the concrete as well as geopolymer as described in the successive sections.

The general properties of all four types of recycled wastes are shown in Tables 4.2 and 4.3, respectively.

FIGURE 4.6   Granules of waste flexible HDPE pipes.

FIGURE 4.7   Recycled palstic wires.

TABLE 4.2

Properties of the Recycled Plastic Wastes in Powder and Granule Forms

Properties	PVC Powder	Plastic Granules	HDP Granules
Specific gravity	1.3	0.7	0.8
Abortion	<0.3	<0.3	<0.2
Color	White	Gray	Dark brown
Shape	Circular	Irregular	Rectangular
Surface	Smooth	Rough	Smooth

TABLE 4.3
Properties of the Recycled Plastic Wires as Fibers

Test Properties	Test Method	Test Value
Material identification	FTIR spectroscopy	Polypropylene – mineral Filled
Tensile breaking load	At 100 mm/min	76 N
Density	ASTM D 792	0.97 g/cm^3

The purpose of the section is to discuss the importance of the awareness of the properties of the waste to be utilized as alternative material in concrete. There are three different forms of waste, which means the constituents to be replaced will also be in a similar form. Therefore, the powder of the recycled plastic was partially added into the cement proportion, the coarse natural aggregates were replaced by the flakes, and sand was replaced by the recycled granules. One important parameter was the surface texture. As can be observed from the table, the plastic granules or flakes had a rough surface while the powder and HDP granules had smooth surfaces. The intermolecular bonding of the other ingredients in the matrix of concrete will also depend on the texture of the aggregates. Therefore, selection of the waste, especially in the case of plastic wastes, should be carried out carefully and with enough study. The second parameter is the density of the material. The plastic granules and powder showed reduced density in the range of 510–750 kg/m^3 and that is nearly half of the conventional cement and natural sand or aggregates. Now this factor will influence the density or self-weight of the overall concrete being produced. Also, the reduced weight of the mix will cause challenges of workability, compaction ability, and strength to an extent. Therefore, care should be taken before using the waste as there may be a significant adverse effect on the overall performance of the matrix or composite for the intended use.

4.2 PROPERTIES OF INDUSTRIAL WASTES IN CONCRETE MANUFACTURING

In this section, details regarding industrial waste are discussed. For the pilot study and to make the selection process effective, there were two types of concrete composites proposed: (a) cement-based concrete and (b) geopolymer-based concrete. The latter concrete is a potential matrix to accommodate significant amounts of industrial waste. For the exploration, pilot studies have been carried out on both types of mixes. The pilot study and results are discussed in the preceding section. Here in this section let us explore some important properties of the waste materials generated from the industry.

4.2.1 FLY ASH

One of the extensively used industry-based waste materials is fly ash. Fly ash, as explained earlier, is a waste produced during coal or similar fuel burning in the

TABLE 4.4
Fly Ash Chemical Composition

Oxide	Class F (%)	Class C (%)
Silica (SiO_2)	50	57
Alumina (Al_2O_3)	28	15
Ferric oxide (Fe_2O_3)	12	4
Calcium oxide (CaO)	6.5	12
Magnesium oxide (MgO)	06	1.1
Potassium oxide (K_2O)	1.5	2
Sodium oxide (Na_2O)	0.2	0.5
Titanium dioxide (TiO_2)	0.1	0.2

generation of electricity at thermal power stations. Depending on the coal types and fuel types, such ashes are categorized as class F and class C fly ash. Class F fly ash are useful in concrete making mixtures. It is rich in silica and alumina oxide contents. The details of chemical compositions are described in Table 4.4. Along with the chemical composition, the relevant physical and mechanical properties have also been discussed. Another reason for the popular utilization of fly ash in concrete is the particle size and the ability of the material to exhibit uniformity and coherence with the other constituents in composite.

From Table 4.4, it is evident that fly ash is a silica-rich material. Moreover, it consists of a good proportion of alumina as well. Therefore, it is usually considered as a pozzolanic mineral material that can participate in gel formation conveniently. Upon sufficient heat provided to the raw form of fly ash, it becomes active and participates in the hydration process of the gel formation also. This quality makes it suitable for creating cement-free concrete. By comparing the chemical ingredients of both types of ashes, the class C fly ash excels in calcium oxide and other similar oxides, namely magnesium and potassium oxide. This makes the fly ash of class C more exothermic while interacting with water during the mixture preparation. Another consequence of adding class C fly ash is that it creates more initial cracks in the hardened mass of the matrix referred as the drying shrinkage cracks. Moreover, the inter-transition zones at the paste and aggregates are also susceptible to a weak adhesion as the alkali silica action also gets a boost with the presence of calcium and similar oxides in the fly ash of class C. Though the classes are divided into two different categories, their particle sizes remain nearly the same or within the same range to an extent as shown in Table 4.5.

4.2.2 FOUNDRY SAND

The manufacturing industrial units have been generating specific types of waste referred to as the used foundry sand or simply foundry sand. Largely the source of this waste is the casting industries. The average production of waste sand from foundries is nearly 100 million metric tons. The quantity has been scaled up in

TABLE 4.5
Fly Ash Particle Size Distribution

Grain Size	Retention (%)
4 mm	9
2 mm	21
500 μm	21
250 μm	24
125 μm	16
90 μm	9
Total	100

the past few years with the advanced technology in casting and metal industries. Foundry sand is the sand chemically bonded to form molds for shaping the molten metals for machines and tool parts. The interesting part is the same foundry sand can be repeatedly used for a couple of times in casting process but ultimately it becomes waste and is disposed into the open streams of the environments, namely land areas. Efforts have been made to utilize the waste foundry sand in concrete synthesis as partial replacement of the fine aggregates or the conventional fine sand. Depending upon its use in the industry, the foundry sands are classified into two primary types as mentioned below.

Two general types of binder systems are used in metal casting depending upon which the foundry sands are classified as: clay bonded systems (Greensand) and chemically bonded systems. Both types of sands are suitable for beneficial use, but they have different physical and environmental characteristics.

Green sand molds are used to produce about 90% of casting volume in the U.S. Green sand is composed of naturally occurring materials that are blended: high-quality silica sand (85%–95%), bentonite clay (4%–10%) as a binder, a carbonaceous additive (2%–10%) to improve the casting surface finish, and water (2%–5%). Green sand is the most used recycled foundry sand for beneficial reuse. It is black in color, due to carbon content, has a clay content that results in a percentage of material that passes through a 200 sieve, and adheres together due to clay and water.

Chemically bonded sand is used both in core making where high strengths are necessary to withstand the heat of molten metal, and in mold making. Most chemical binder systems consist of an organic binder that is activated by a catalyst, although some systems use inorganic binders. Chemically bonded sands are generally light in color and in texture than clay bonded sands (Figure 4.8).

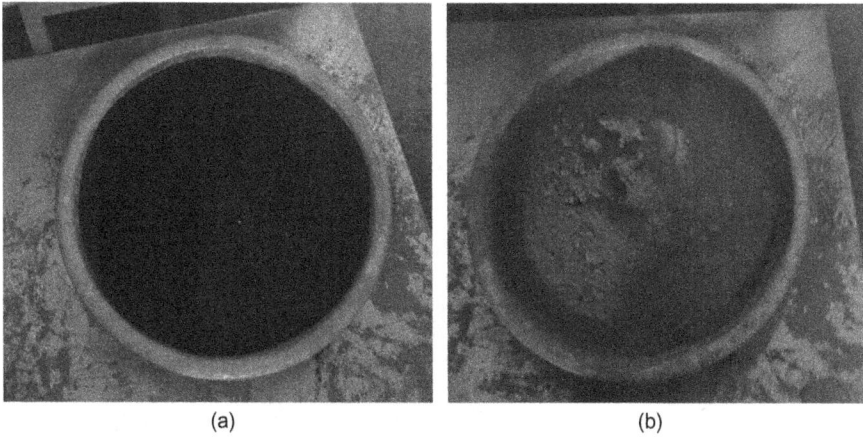

(a)   (b)

FIGURE 4.8   (a) Green foundry sand. (b) Chemically bonded foundry sand.

4.2.3   PROPERTIES OF FOUNDRY SAND

4.2.3.1   General and Chemical Properties

Foundry sand is typically subangular to round. After being used in the foundry process, a significant number of sand agglomerations are formed. When these are broken down, the shape of individual sand grains is apparent. Typical physical properties of spent foundry sand from green sand systems are given in Table 4.6. The grain size distribution of spent foundry sand is very uniform, with approximately 85%–95% of the material between 0.6 and 0.15 mm (No. 30 and No. 100) sieve sizes. About 5% to 12% of foundry sand can be expected to be smaller than 0.075 mm (No. 200 sieve). The particle shape is typically subangular to round. Waste foundry sand gradations have been found to be too fine to satisfy some specifications for fine aggregate. Table 4.7 indicates the chemical properties of foundry sand.

TABLE 4.6
General Properties of the Used Foundry Sand

Specification/Property	Values
Specific gravity	2.2–2.5
Bulk relative density (kg/m³)	2,540
Absorption (%)	0.45
Moisture content (%)	0.15–9
Clay lumps	1–40
Coefficient of permeability (cm/s)	10^{-3}
Plastic limit	Non-plastic

TABLE 4.7
Chemical Properties of Used Foundry Sand

Chemical Component	Proportion in %
SiO_2	87.91
Al_2O_3	4.7
Fe_2O_3	0.94
CaO	0.14
MgO	0.3
SO_3	0.09
Na_2O_3	0.19
K_2O	0.25
TiO_2	0.15
SrO	0.03
LOI	5.15

The chemical composition of foundry sand relates directly to the metal molded at the foundry. This determines the binder that was used, as well as the combustible additives. Typically, there is some variation in the chemical composition of foundry sand from foundry to foundry. Sands produced by a single foundry, however, will not likely show significant variation over time. Moreover, blended sands produced by consortia of foundries often produce consistent sands. The chemical composition of the foundry sand can impact its performance. Spent foundry sand consists primarily of silica sand, coated with a thin film of burnt carbon, residual binder (bentonite, sea coal, resins), and dust.

4.2.3.2 Mechanical Properties

Spent foundry sand has good durability characteristics as measured by low Micro-Deval abrasion and magnesium sulfate soundness loss tests. The Micro-Deval abrasion test is an attrition/abrasion test where a sample of the fine aggregate is placed in a stainless steel jar with water and steel bearings and rotated at 100 rpm for 15 minutes. The percentage loss has been determined to correlate very well with magnesium sulfate soundness and other physical properties. Recent studies have reported relatively high soundness loss, which is attributed to samples of bound sand loss and not a breakdown of individual sand particles. The angle of shearing resistance (friction angle) of foundry sand has been reported to be in the range of 33°–40°, which is comparable to that of conventional sands.

Used foundry sand is a partial replacement of cement or a partial replacement of fine aggregates or a total replacement of fine aggregate and a supplementary addition to achieve different properties of concrete. By using foundry sand, we save the cost of original materials and make low-cost concrete. Using foundry sand led to the development of environmentally friendly construction. As scientific research shifts focus to smaller scales, new and innovative research methods are needed to develop new concrete. Utilization of waste in concrete pavement/structures will give us a longer life span and minimize maintenance.

4.2.4   FURNACE SLAG AND SILICA FUMES

Ground granulated blast furnace slag (GGBS) is a by-product of manufacturing iron, while iron ore, limestone, and coke are heated to 1,500°C in the blast furnace. When these materials are melted in a blast furnace, two products are produced – hot metal and slag. The slag is lighter and floats on top of molten iron. Molten slag mainly includes silicate and alumina from iron ore, and some oxides from limestone. Process of granulating the slag includes cooling the slag through a high-pressure water jet. This causes the slag to rapidly quench and form granular particles with maximum diameter of 5 mm. Rapid cooling prevents the formation of greater crystals and the resulting granular material comprises about 95% amorphous calcium alumina-silicate. Drying and then grinding into a finer powder produce GGBS cement. The grinding of the granulated slag is carried out in a rotary ball mill. GGBS is mainly composed of CaO, SiO_2, Al_2O_3, and MgO. It has the same main chemical composition as ordinary Portland cement, but in different proportions. Table 4.8 illustrates the general properties of furnace slag.

The furnace slags are considered as one of the richest sources of calcium and silica oxides like the class C fly ash. The chemical composition is shown in Table 4.9.

TABLE 4.8
General Properties of Furnace Slag (GGBFS)

Property	Value
Physical form	Off white powder
Bulk density (kg/m³)	1,200
Specific gravity	2.9
Specific surface (m²/kg)	425–470

TABLE 4.9
General Chemical Composition of the Furnace Slag Powder

Oxide	Percentage (%)
Silica (SiO_2)	36.7
Alumina (Al_2O_3)	17.20
Ferric oxide (Fe_2O_3)	1
Calcium oxide (CaO)	34.62
Magnesium oxide (MgO)	8.9

FIGURE 4.9 Powder form of the grounded blast furnace slag.

4.2.5 SILICA FUMES

Silica fume, also known as micro silica powder, is an amorphous polymorph of silica. It is an ultrafine powder collected as a by-product from the production of silicon and ferrosilicon alloy and is composed of spherical particles having an average particle diameter of 150 nm. The main application area is the high-performance concrete volcanic ash material. Silicon powder is an ultrafine material having spherical particles with a diameter of less than 1 μm and an average of about 0.15 μm. This makes it about 100 times smaller than the average cement particle. The bulk density of silica fume depends on the degree of densification in the silo, ranging from 130 (not densified) to 600 kg/m^3. The specific gravity of silica fume is usually in the range of 2.2–2.3. This is as per IS 15388 and ASTM C 1240, "standard specification of silica fume as mineral admixture in hydraulic cement concrete and mortar."

The chemical composition of the silica fume is indicated in Table 4.10. It may be noticed that the composition of the silica fume shows excellent content and proportions of silica in the form of silicon oxides; therefore, the powder is readily involved in chemical activities with the other chemical compounds to form a gel-like structure and also contribute to the heat generation necessary for the polymerization process, especially in the case of cementless concrete or geopolymer concrete.

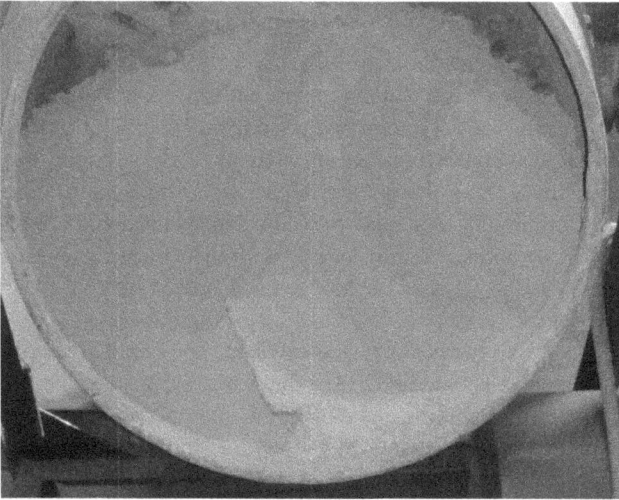

FIGURE 4.10   Powder form of the silica fume.

TABLE 4.10
General Chemical Composition of Silica Fume Powder

Oxide	Proportion in %
Silica (SiO$_2$)	92.8
Alumina (Al$_2$O$_3$)	0.6
Ferric oxide (Fe$_2$O$_3$)	0.3
Magnesium oxide (MgO)	0.6

The description of the furnace slag and silica fume indicates that the materials are important constituents for making cement-free concrete. The availability of the oxides in the material creates strong and stable bonds with water and other admixtures or chemicals for making a matrix. The essential role of the silica fume is to provide energy to the matrix for the formation of the gel or binder constituents and making the overall mix capable of obtaining a strong hardened state. This is significant because the conventional method of making geopolymers requires oven curing or external heat sources for initiating polymerization in the fresh mix. However, it may be more energy intensive and not practically possible for the members of larger size and scale. Though the option of steam curing is available for the larger sized members, the bonding is difficult to ensure for the entire matrix. The presence of silica fume eliminates the requirement of any external heat for the fresh mix and provides the hardened and strong physical form of the concrete at the ambient temperature and also without of the restriction of the scale of the members. The utilization of these materials is illustrated in the preceding sections of the pilot studies.

4.2.6 Chemicals and Polymeric Compounds

Nowadays, several chemical admixtures are in practice for making concrete and concrete-like composites with and without the use of cement as a binder material. There are specific chemicals and polymeric compounds prepared to perform specific tasks in the composite. The admixtures are utilized for early strength, long-term durability, water resistance, rapid hardening, higher intermolecular bonding, and many other desired outcomes. In this section, a few important details have been shared about sodium silicate and sodium hydroxide with their illustrative usage in composite preparation as well as latex as the source of increased internal bonding agent for concrete constituents.

Sodium silicate is a chemical capable of initiating chemical reaction within the composite constituents. The addition of sodium hydroxide plays the role of energy-supplying compound for breaking and joining the internal bonding of the oxides available in the composite. When added they are referred as the alkaline solutions for concrete. Sodium-based solutions were chosen because they were cheaper than potassium-based solutions. Technical-grade sodium hydroxide solids were used in flake form (3 mm), with a specific gravity of 2.21 and 98% together. The sodium hydroxide (NaOH) solution was prepared by dissolving either the flakes or the pellets in water. The mass of NaOH solids in a solution varied depending on the concentration of the solution expressed in terms of molar, M. For instance, a NaOH solution with a concentration of $8\,M$ consisted of $8 \times 40 = 320$ g of NaOH solids (in flake or pellet form) per liter of the solution, where 40 is the molecular weight of NaOH. In this experimental work, sodium hydroxide was used in three different concentrations of 8, 12, and 16 M.

Another category of admixture is polymer. Polymers play a vital role in the enhancement of the material properties of concrete and concrete-like composites. The polymer-modified concrete is a primary source for the development of modern concretes. One such most used polymer compound is latex. Latex can be produced in laboratories and is available in nature. However, the natural latex should be refined and conditioned for appropriate property enhancement. The primary role of the latex in concrete is to create better adhesion and bonding of the constituents. It is also sometimes known as synthetic rubber. It is a water-based emulsion of styrene butadiene copolymer particles. The properties of the latex are shown in Table 4.11.

(a)   (b)

FIGURE 4.11   (a) Sodium hydroxide flakes. (b) Sodium silicate liquid.

TABLE 4.11

General Properties of Latex Compound

Properties	Results
Appearance	Free flowing liquid
Color	Milky white
Specific gravity @ 30°C g/mL	$1.02 + 0.02$
Nonvolatile matter, %	42–44
Bond strength, N/mm^2	5
pH value	7–9
Chemical resistance	Resist mild acids and alkalis
Freeze–thaw resistance	Excellent

4.2.7 MINING WASTES

Mining and quarrying activities are carried out in nearly all the countries and are also necessary to obtain useful minerals and materials. Construction is a sector that remains largely dependent on mine-based materials, namely sand, aggregates, limestones, marbles, special clays, and many mineral-based materials. In this section, a few important and commonly used mining wastes, namely marble wastes, are discussed for their properties and utilization scenario.

Marble is obtained in varying forms, colors, and quality from marble stone mines. India produces around 15 million tons of marble every year and majority are produced in the state of Rajasthan. Marble and marble products are used in many ways in construction. The demand has been increasing and the mining activities have also been faster. However, the mining process, namely cutting, shaping, and polishing, generates varying types of waste in a huge quantity on a daily basis. The primary waste types include marble dust, marble sludge, and marble chips. Though the wastes are useful for their secondary applications, the waste quantity is very huge and difficult to handle for reuse and largely dumped into the natural streams, resulting in pollution. Marble varies in their properties and chemical compositions though; Table 4.12

TABLE 4.12

General Properties of Marble Sludge/Dust

Fineness	3
Brightness	92
Retention on 90 µm sieve	10%
Moisture	49%
Acid solubility	2%
Specific gravity	2.7
Hardness	3%
Liquid limit	18%
Shrinkage limit	23%

describes the general properties and common chemical composition of the waste from marble mines. The figure illustrates the marble waste generation scenario.

The marble waste has been observed in many civil or construction-related applications nowadays. The major uses are listed below:

- Plasters and coating
- Ornamental and external surfacing work
- Slurry for water proofing treatments
- Mortar making for masonry construction
- Casting of small- to medium-scale pre-cast nonstructural members
- Filler material in lean concretes
- Making of paints, pigments, and powder for coating

Over and above these applications, researchers have practiced and explored the possibilities to develop concrete using marble wastes by partially replacing the conventional materials, namely cement and fine sand in concrete. The results showed that while replacing cement in a mix, it is necessary to utilize the admixtures supporting bond strength enhancement as marble can be a pozzolanic material but without binding capacity. The standards on the concrete mix design, however, have mentioned that the addition of pozzolanic material of 30% may be explored for possible inclusion in concrete. However, the test results should be referred from the literature before implementing the proportion in practice.

The list of industrial waste is in fact enormously long and with modern materials, there are several options available, though the use should be based on pilot studies.

4.3   PILOT STUDIES AND PRELIMINARY EXPERIMENTAL EVALUATION OF MODIFIED CONCRETE

The utilization of any waste material in concrete requires significant detailed study on primarily two attributes. One is analyzing the properties of the wastes to determine their suitability for replacement with conventional constituents and second is the investigation of the mixtures for their response to how the material containing the wastes reacts. Therefore, pilot studies on the concrete containing the waste plastics and the other industrial wastes are highly recommended. In this section a few of the pilot studies have been discussed. This section presents the results and discussion on how the addition of plastic waste may influence the behavior and response of conventional concrete. It is learnt to what extent conventional concrete changes its properties and through that a decision may be made and feasibility of addition of the waste may be finalized for detailed evaluation of the mixtures. The aim of the pilot study was to fix parameters influencing the results of experimental investigations. Primarily three attributes were projected as parameters, namely water to cement ratio for concrete mixes, fractions of metallized plastics to be added in concrete, and dimensions of shredded metallized plastic. These attributes were fixed as parameters by investigating basic workability and strength properties on trial mixes.

4.3.1 WATER–CEMENT RATIO

Strength and durability properties of concrete are affected by cement contents, and availability of water in the mix is one of the important attributes. Three different values of water to cement ratios, i.e., 0.45, 0.55, and 0.65, were adopted to prepare concrete mixes. The value of 0.65 for the water to cement ratio was included to cover the response of the concrete containing PCMPW for nonstructural applications.

4.3.2 PCMPW FRACTIONS AND DIMENSIONS

Inclusion of waste plastic as a constituent in concrete resembles fiber-reinforced concrete (FRC). Literature review showed that fibers have been added in concrete mixes by volume fractions of the mixes. Waste plastics, namely PET bottle waste and bags, have been rendered into fibers and chips of varying sizes and added into concrete in varying fractions.

For the present work, metallized plastic sheets were converted into flakes and fiber form and added by varying volume fractions.

Unlike the rigid form of PET waste plastics, thin and smooth metallized plastic films were found difficult to cut in a regular micro size and shape and even difficult to produce in a fiber form. Therefore, shredding was done in different macro sizes in the form of flakes like 1, 5, and 10 mm as average sizes with a constant thickness. The metallized plastic sheets were manually cut into macro fiber form of 1 mm × 10 mm and 1 mm × 20 mm fibers as width and length dimensions respectively.

A review of literature available on the usage of waste plastic in concrete revealed that irrespective of the dimensional properties of plastic wastes, viz. macro size, micro size, fibers, flakes, chips, or granules, the addition was done in very small fractional values. Waste plastic is added in small dosages like 0%, 0.5%, 1%, etc. by volume of the concrete mixes. For the pilot study, all flakes and fibers of PCMPW were mixed in small fractions ranging from 0% up to 5%. The primary purpose of the pilot study was to obtain the most significant sizes and fractions of PCMPW to be added in the concrete mixes throughout the study. Preliminary experimental investigations were carried out on fresh and hardened concrete containing PCMPW. The objective of the pilot study was to select the most suitable test parameter values to be included for the full-length investigations. A total of three trial mixes were prepared with varying water to cement ratio of 0.45, 0.55, and 0.65. PCMPW was added in each mix with five varying sizes and fraction range from 0% to 5%.

4.3.3 FRESH PROPERTIES

Slump test was conducted on freshly mixed concrete containing PCMPW in shredded form of varying dimensions and fractions from 0% to 5% by volume of concrete. PCMPW affected the homogeneity of concrete. Plastic particles tend to occupy water-cement paste around them and reduced adequate viscosity in the mix to generate slump suitable to workable concrete.

Slump values were influenced by the size of PCMPW. PCMPW shredded into flakes of average size of 10 mm made concrete highly non-workable from the time of mixing. A sudden loss of viscosity in the mixes was observed. Moreover, the large-sized flakes interrupted cohesion between aggregates and cement paste while performing the tests. A similar response was exhibited by PCMPW cut into 1 mm × 10 mm fiber form. The fibers with length of 10 mm were found incapable to act as fibers owing to the small length compared to their width of 1 mm. It was finally concluded not to use PCMPW with 10 mm flakes and 1 mm × 10 mm fibers.

Increased fraction of PCMPW reduced slumps for all sizes and forms. However, beyond the dosage of 1% by volume of concrete, except 1 and 5 mm size flakes and 1 mm × 20 mm size fibers, there was significant loss of slump. The full dosage of PCMPW, namely 5% by volume of concrete, made the mixes very stiff and difficult to work. Therefore, 5% dosage of PCMPW was found unfit for the full-length experimental study.

4.3.4 STRENGTH PROPERTIES

The addition of PCMPW reduced the compressive strength of concrete. The strength was influenced by PCMPW size and fraction. Large-sized flakes of average 10 mm size and short fiber of 1 mm × 10 mm significantly reduced the strength up to 30% at a dosage of merely 1% by volume of concrete. The strength continued to reduce with increased fraction beyond 1% dosage and reduced up to 60% at the dosage of 2% and up to 75% at 5% dosage of PCMPW of 10 mm flakes and 1 mm × 10 mm fibers.

The pilot study consisting of tests for workability and on one of the primary strengths namely compressive strength of concrete provided important observations that PCMPW in the form of flakes of 10 mm average size and in the form of fibers of 1 mm × 10 mm size should be avoided as they drastically reduced the properties and may not be feasible for use in concrete as a constituent.

The workability and strength property were primarily influenced by PCMPW size and fraction. However, for varying water to cement ratio, both the properties showed constant reduction for the other two parameters. Concrete prepared with 0.65 water to cement ratio has not been preferred for reinforced concrete construction though it was included in the investigations as such concrete are used in many nonstructural components namely leveling, filling, lining, and small structures like poles, dividers etc.

For the complete experimental study, three sizes of PCMPW of average sized flakes of 1 mm, 5 mm, and fibers of 1 mm × 20 mm were selected. The fraction range for the dosage of PCMPW was fixed from 0% to 2%. The water to cement ratio values were, however, not minimized and kept as is with three options, namely 0.45, 0.55, and 0.65.

4.3.5 MIX PROPORTIONS AND BATCHES

Three mixes were prepared with varying water to cement ratios of 0.45, 0.55, and 0.65. The mix design conformed with IS: 1262:2007. The mixed proportions are shown in Table 4.13. Table 4.14 shows mixes and batches prepared for the experimental study.

TABLE 4.13
Mix Proportions for 1 m³ Concrete

Mixture	Cement	Aggregates 20 mm	Aggregates 10 mm	Sand	Water	W/C ratio
	kg	kg	kg	kg	kg	
1	420	670	440	640	185	0.45
2	355	675	450	695	194	0.55
3	305	660	445	715	198	0.65

TABLE 4.14
Batch Designations and PCMPW Details

Water to Cement Ratio	PCMPW Fiber Type	Batch designation	PCMPW fiber fraction
0.45	A	B1 to B5	0% to 2%
	(1 mm)	B16 to B20	0% to 2%
		B31 to B35	0% to 2%
0.55	B	B6 to B10	0% to 2%
	(5 mm)	B21 to B25	0% to 2%
		B36 to B40	0% to 2%
0.65	C	B11 to B15	0% to 2%
	(20 mm)	B26 to B30	0% to 2%
		B41 to B45	0% to 2%

4.3.6 ASSESSMENT OF MATERIAL PERFORMANCE

Material performance was assessed for the alteration in standard property values due to the inclusion of PCMPW in conventional concrete. Tests for workability, strength, durability, and structural response were performed compartmentally in laboratories. Standards relevant to the type of tests were referred to including IS codes and ASTM. Fresh behavior was assessed based on slump and compaction factor test values. Hardened concrete molded into cube, cylinder, disk, and beam specimens was tested to obtain compressive strength, splitting tensile strength, impact strength, and pull-off strength. Impact resistance was examined by drop weight method. Pull-off strength or the bond strength was determined with standard guidelines of ASTM C 1583-04. Resistance to acid, sulfate, and chloride ingress in concrete containing varying fraction and sizes of PCMPW was assessed in accordance with standard guidelines of ASTM codes respectively. Disk specimens were submerged in water to study the rate of water absorption by concrete containing PCMPW flakes.

Experimental values obtained for compressive strength and splitting tensile strength were compared to obtain analytical interrelationship of the quantities. Moreover, the stress and corresponding strain values obtained by the tests on cylinder specimens were used to obtain the interrelationship. Reinforced beam specimens were subjected to flexure to obtain the moment resisting capacity and alterations

TABLE 4.15
Details of Test Specimens

Test	Specimen	Dimensions in mm			Total Numbers
		L	B	H/D	
Compressive strength	Cube	150	150	150	171
Pull-off strength	Cube	150	150	150	45
Acid resistance	Cube	150	150	150	45
Sulfate resistance	Cube	150	150	150	45
Splitting tensile strength	Cylinder		150	300	135
Stress–strain relationship	Cylinder		150	300	12
Impact strength	Disk		150	60	45
RCPT	Disk		100	60	45
Oxygen permeability	Disk		100	60	45
Water sorptivity	Disk		100	60	45
Flexure response	Beam	1,500	150	200	12

due to the PCMPW inclusion. Analytical and experimental results were compared for obtaining the response of beams in flexure as structural application of proposed material. Table 4.15 shows details of specimens prepared for testing.

4.3.7 FRESH CONCRETE TESTS

4.3.7.1 Slump Test

The workability of fresh mix of concrete containing PCMPW flakes in varying proportions was measured by slump tests. Slump behavior indicates stiffness of fresh mix to work with. Factors like size and shape of aggregates and water to cement ratio primarily affect the slump behavior. Slumps are measured as high, medium, and low. Slump values were obtained for every mix before further usage in testing of properties. Slump is a variation in height of the fresh concrete held in the set position by a conical mold with reference to the standard values. Table 4.16 shows standard slump values and usage of concrete.

Slump is maintained according to the need of application. However, slump must be controlled to keep minimum workability of fresh mix. Concrete reinforced with PCMPW exhibited different values for slump for varying sizes and fractions.

TABLE 4.16
Standard Reference Values for Slump

Slump Value in mm	Category	Application or Use
100–150	High	Pumping, piling, trench-fill
50–100	Medium	Primary structural elements
25–75	Low	Mass concrete, pavements
Less than 25	Low	Shallow foundations

The water to cement ratio remained almost standard as the low values reduced slump and high values increased slump.

4.3.7.2 Observatory Notes

Slump of concrete was found to be sensitive to inclusion of PCMPW. Slump response varied with the test parameters. While preparing the concrete for slump test, it was observed that the flakes tend to float on the surface of mix and showed no adherence with constituents at first stage; hence, more time was required for mixing the constituents. With time and number of revolutions, the flakes started getting mixed with the constituents. The mix became stiffer on increased contents of PCMPW. Mix containing types B and C showed more stiffness than type A flakes. The results were recorded for all mixes and compared with slump values of concrete containing 0% PCMPW flakes.

Effects of type of PCMPW on slump were observed carefully. Type B showed more reduction in slump compared to other types. Inclusion of higher volume fractions of flakes resulted in non-uniform mixes. Flakes obstructed cement gel formation and increased balling effects, resulting in segregation of aggregates from the matrix. However, small sized flakes remained ineffective on mixing response of constituents at dosage up to 1%. A similar response was noticed in case of type C of PCMPW.

Slump test are shown in Figure 4.12. The figure shows variation of slump values with respect to PCMPW fractions for three different water–cement ratios.

In all mixes, inclusion of PCMPW in concrete reduced the slump, which is visible from the trend. Slump reduction indicated that addition of PCMPW makes concrete stiff and less viscous. The viscosity of a mix improves spreading of concrete in formworks and reduces air voids in a mix. A concrete may not be workable if slump continues to reduce for varying fractions of PCMPW. Therefore, it was required to obtain the optimum PCMPW dosage and type.

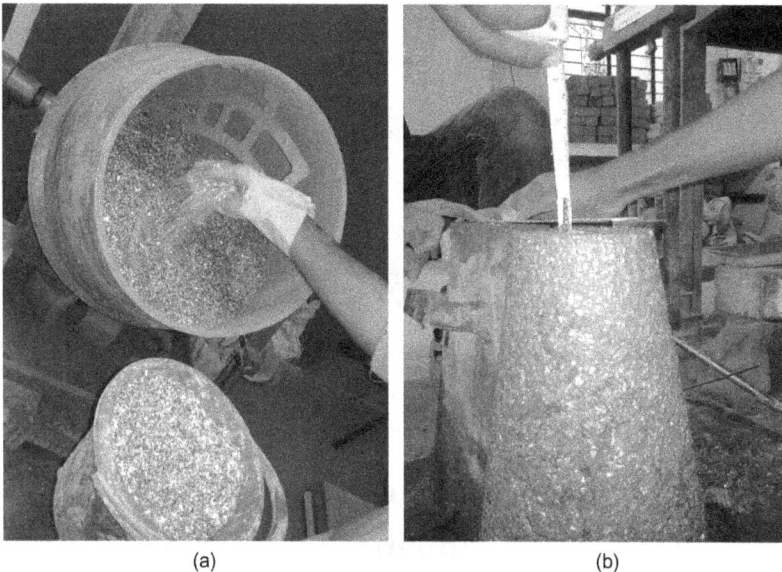

(a)   (b)

FIGURE 4.12   (a) Dry mix of PCMPW in concrete. (b) Slump test images.

Results showed that water to cement ratio did not affect the trend of reduction of slump for given PCMPW attributes, namely fraction and type. According to the test results, the most significant parameter for slump response was the fraction of PCMPW. Moreover, the trends of slump reduction were similar for concretes containing types A and C of PCMPW and type B showed larger reduction of slump relatively.

The effect of fraction addition of 0.5% of PCMPW of types A and C reduced slump up to 4% and continued to reduce up to 7%, 12%, 14%, and 21% for every 0.5% increment in fraction of PCMPW to 2% by volume of concrete. Type B reduced slump up to 34% of addition at 2% addition by volume of concrete. Slump results exhibited relatively less reduction for types A and C PCMPW up to 1% addition by volume of concrete. For further details kindly refer the work published by the author [1].

PCMPW of type B was found responsible for increased loss of viscosity by restricting homogeneous mixing of constituents. Larger size of flakes such as type B plastic increased segregation of aggregates from cement paste and increased stiffness of the mix. The results showed that the optimum dosage of PCMPW should be 1% with type A or type C PCMPW to be used in concrete to maintain its workability (Figure 4.13).

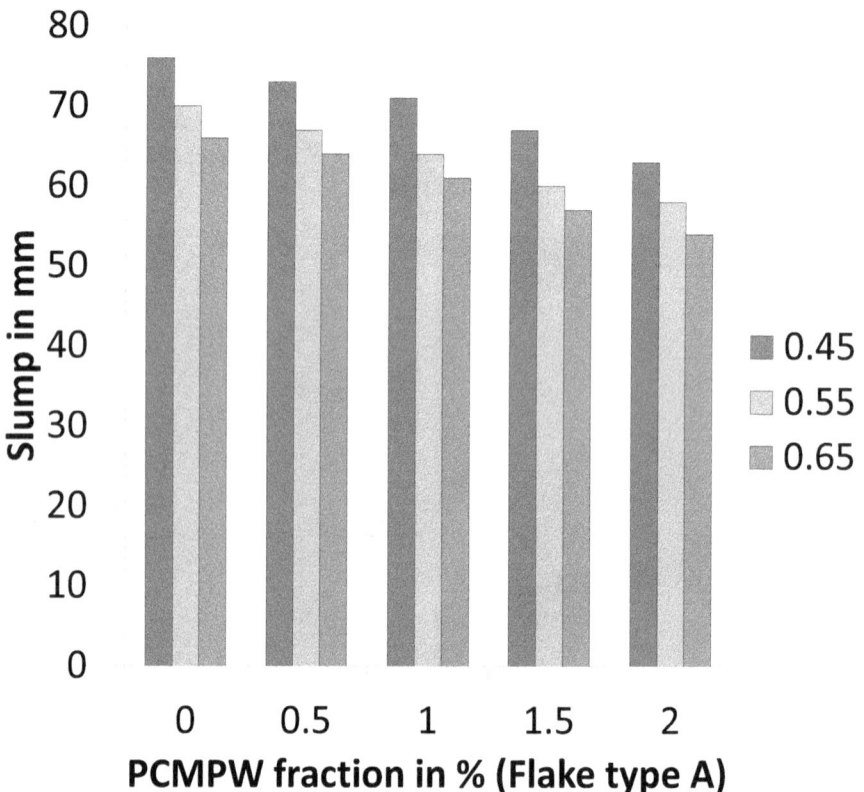

FIGURE 4.13   Effect of PCMPW fraction in slump of concrete.

TABLE 4.17
Standard Reference Values of Compacting Factor

Compacting Factor Test	Category	Application or Use
0.95	High	Pumping, piling, trench-fill
0.92	Medium	Primary structural elements
0.85	Low	Mass concrete, pavements
0.78	Very low	Shallow foundations

4.3.7.3 Compacting Factor

Compacting factor test is required when the concrete is relatively stiff and shows less slump (IS: 1199-1959). The test is sensitive for workability measurement and response as it accounts for the effects of initial hydration process of cement in concrete. Freshly mixed concrete should be promptly used for the test to obtain accuracy in results. Like slump values, Table 4.17 shows ranges of compacting factors according to their usage.

4.3.7.4 Observatory Notes

Conforming to the requirements of IS: 1199-1959 codes, standard apparatus was used for obtaining compacting factor values of each mix. The time of shutter release was strictly followed not more than 2 minutes between the two consecutive hoppers. It was observed that concrete containing type B flakes showed less free flow while allowing the drop of concrete from the upper hopper to the lower hopper. Figure 4.14 shows a compacting factor test being carried out.

FIGURE 4.14 Compacting factor test.

Compaction factor test results are shown in Figure 4.15. The trend lines showed a gentle reduction in values with increased PCMPW fractions in the range of 0%–2% by volume of the mix. The results represented in graphical format revealed very little variation in compaction factor values by varying test parameters.

Concrete containing types A and C of PCMPW showed almost similar and gradual trend of reduction, while concrete containing type B differed in response showing steep reduction trend lines. This scenario was observed to be common for all the three types of concrete. It was observed that concrete prepared with 0.45 as water to cement ratio showed minimum reduction in values of compacting factor among all three mixes.

Compaction factor reduced up to 5% at 0.5% addition of PCMPW by volume of concrete. The values reduced up to 10% at 1% addition of PCMPW and continued to reduce up to 20% at full dosage of PCMPW at 2% by volume of concrete. It was observed that maximum reduction of 22% of values was in case of PCMPW type B for all three types of concrete mixes.

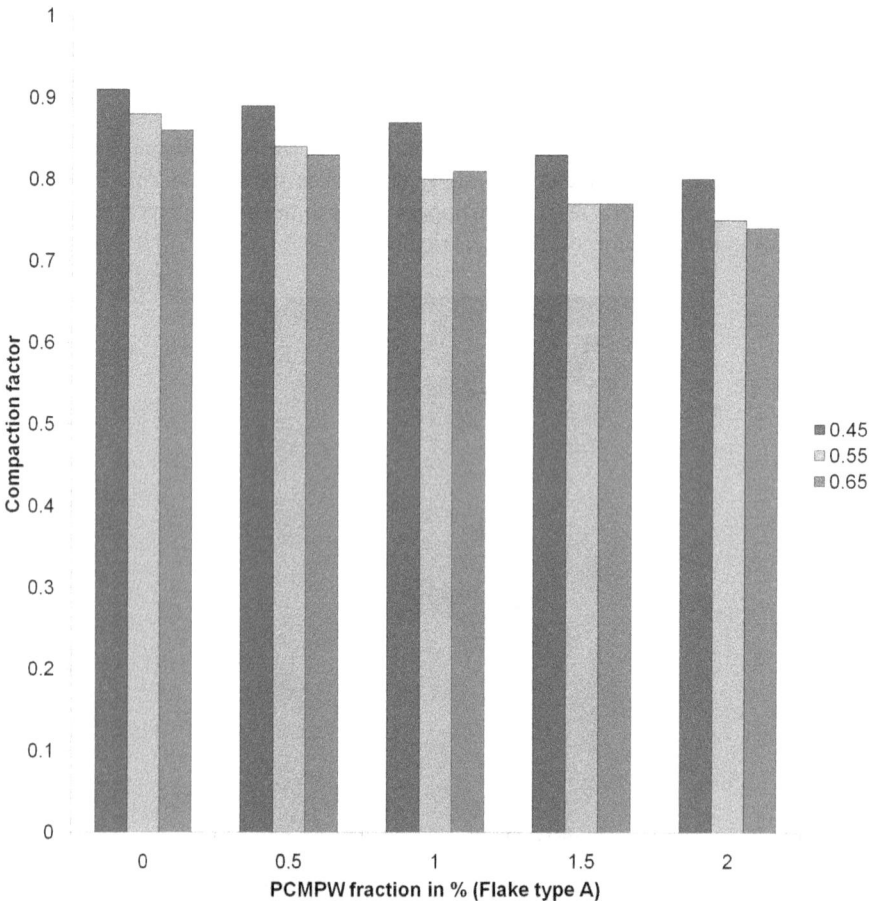

FIGURE 4.15   Effect of PCMPW fractions on compaction factor of concrete.

PCMPW types A and C reduced the values up to 15% as maximum reduction, while type B reduced values up to 22%. At dosage of 2% of PCMPW, types A and C showed minimum reduction of 11% specifically in case of concrete prepared with 0.45 as water to cement ratio.

Compaction factor test results reduced with increased PCMPW fractions. The presence of PCMPW obstructed the rearrangement of aggregates in the mix and expelled entrapped air in the mix. Concrete mixes showed segregation of constituents at higher dosage of PCMPW and reduced degree of compaction. The free falling of concrete mass containing PCMPW between the hoppers was also affected with increased PCMPW fractions.

A careful observation on test results and experience of working with fresh concrete containing PCMPW during the compaction factor tests led to the observation that PCMPW can be added to the concrete with a dosage of 1% by volume of mix as the consistency remained the same at this dosage in all mixes. The preferred dimension of PCMPW particles should be about 1 mm average as flake or 1 mm × 20 mm fiber form.

4.3.8   Effect of PCMPW Fraction

4.3.8.1   Strength Tests

Refer Figure 4.15.

4.3.8.2   Compressive Strength

Cube and cylinder specimen conforming to the requirements of IS: 516-1999 was prepared to obtain strength of concrete in axial compression. A compression testing machine of 2,000 kN capacity was used to determine compressive strength of hardened concrete modified by PCMPW flakes. All molds were oiled and tightened against leakage. Specimens were water cured for 28 days at room temperature. An average value of three tests was recorded. Figure 4.16 shows the arrangement for the same.

4.3.8.3   Observatory Notes

While performing the tests, the reference concrete failed abruptly by showing instantaneous cracks. The response was different in the case of concrete containing varying fractions of PCMPW flakes. Regardless of the water to cement ratio and types of flakes, the concrete containing low fractions of PCMPW, i.e., 1% by volume of concrete, showed uniform micro cracks on the exterior surfaces. At final failure also, the specimen remained bounded and demonstrated ductile failure including bleeding of concrete. More details related to the methods and testing may be referred in the authors' previous publications listed in the references [1-4].

Figure 4.17 shows variation in the compressive strength with the varying PCMPW fractions and types for all the three mixes. The trend lines indicated that the strength reduction scenario was nearly similar for all concrete mixes prepared with different water to cement ratios, namely 0.45, 0.55, and 0.65.

Concrete prepared with water to cement ratio of 0.45 showed minimum strength reduction at a constant fraction and type of PCMPW. The strength reduced as negligible as 3% to the maximum of 18% for the tests conducted on the specimens prepared

FIGURE 4.16   Compressive strength test arrangement and tested specimen.

by mix 1 (w/c: 0.45). However, the strength reduced in the range of 5%–29% for the dosage range of PCMPW from 0.5% to 2% dosage for the same mix in case of type B PCMPW.

Concrete prepared with a water to cement ratio of 0.55 also demonstrated limited reduction of 3%–9% at a lower dosage of 0.5%–1%, respectively. However, unlike mix 1, the strength reduction scenario varied with type of PCMPW and type C of PCMPW reduced the strength in the range of 6%–12% at a lower dosage of 0.5% and 1%, respectively, and showed larger reduction of 24% at final fraction of 2% by volume of concrete.

4.3.8.4   Effect of PCMPW Fraction

Strength reduced according to the types of PCMPW with the increased fractions of PCMPW. Strength reduced in the range of 3%–8% for the addition of PCMPW of 0.5% by volume of concrete. The strength was further reduced in the range of 9%–12% with the increased dosage of PCMPW as 1% by volume of the mix. For the increased PCMPW fractions from 1% to 2% by volume of concrete, the compressive

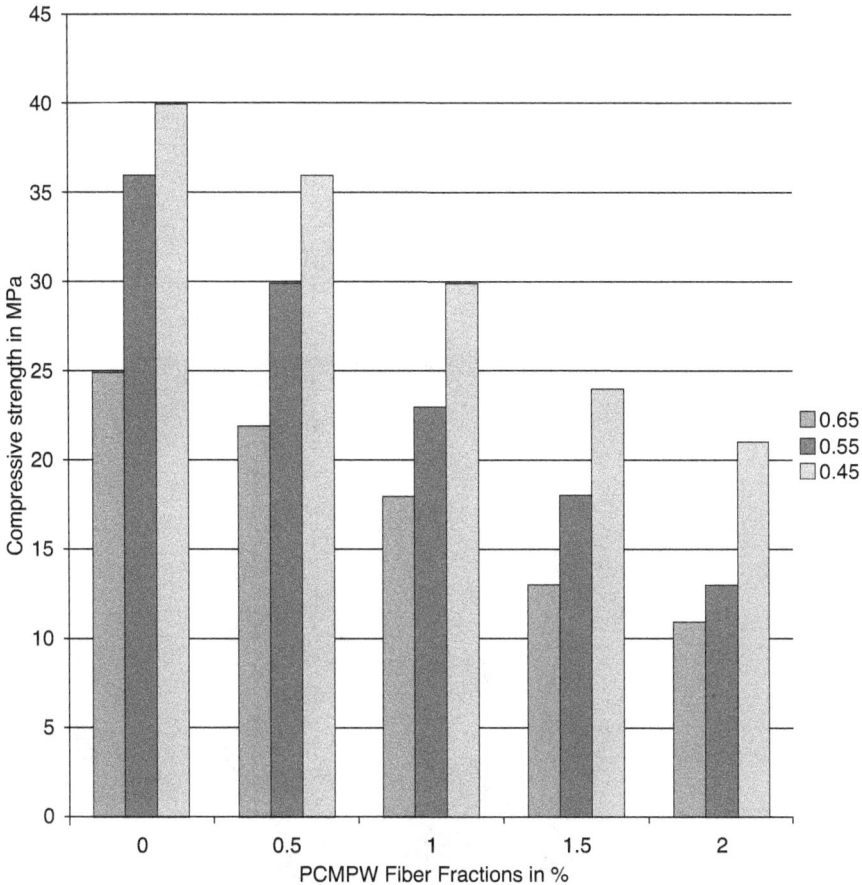

FIGURE 4.17   Effects of PCMPW flakes on compressive strength.

strength reduced in the range of 18%–24%. It is to be noted that the results discussed are for the mixes consisting of PCMPW of types A and C.

4.3.8.5   Splitting Tensile Strength

Cylinders were tested for splitting tensile strength in accordance with IS: 5816-1999 for obtaining strength of concrete under indirect tension. Response of concrete from crack formation at the initial stage and final failure was studied. The test measures crack resisting capacity of concrete by indirect compression applied across the longitudinal axis of the specimen. Splitting tensile strength is expressed by fraction of compressive strength. Resistance of concrete to crack formation depends on compressive strength of concrete but at the transition zone around the aggregates and cement paste molecular adhesion plays a major role. Tensile failure of concrete is a resultant effect of failure of cement paste.

4.3.8.6   Observatory Notes

Crack patterns were carefully observed during the tests. Initial minor cracks across the diameter were gradually widened with increased loading in case of reference concrete. PCMPW affected cracking patterns, especially in the case of types A and C, and was found to be capable of getting grip in hardened mass. Type C showed significant resistance against cracking and was found effective in reducing the crack propagation. Figure 4.18 shows the test arrangement, and Figure 4.19 shows the influence of the addition of PCMPW fibers in concrete with a significant improvement in the resistance of the concrete to the splitting actions.

FIGURE 4.18   Splitting tensile strength test apparatus and specimen.

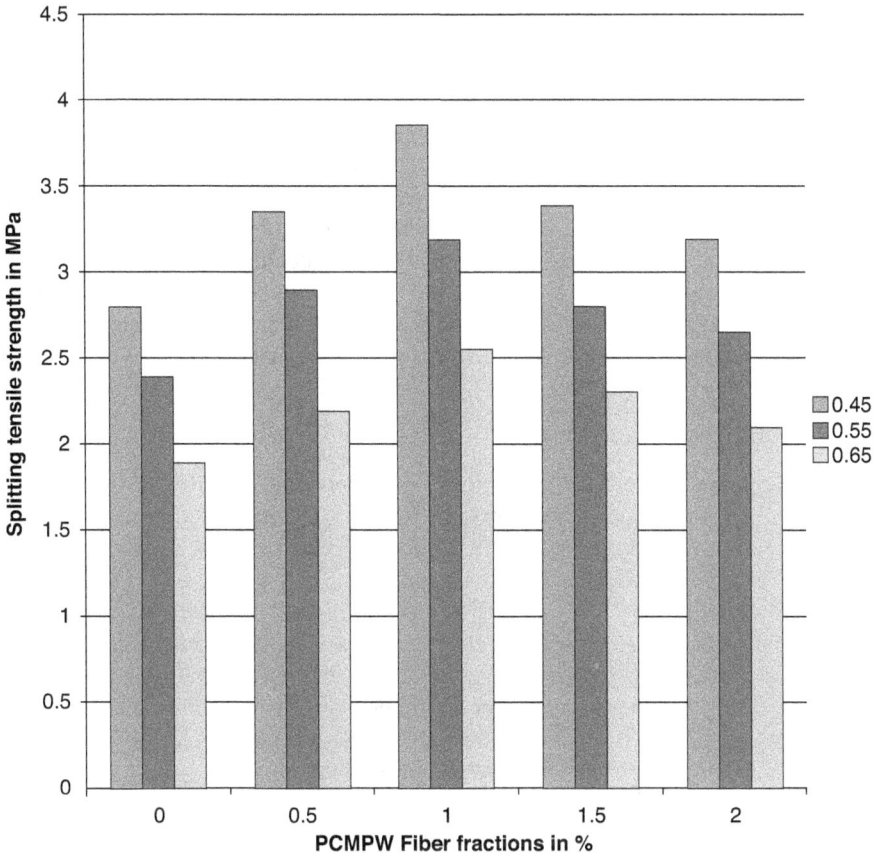

FIGURE 4.19 Influence of the PCMPW fibers on the splitting tensile strength resistance of concrete.

4.3.9 Effects of PCMPW Fraction

Upon addition of PCMPW of 0.5% dosage, the strength improved in the range of 10%–14% in case of types A and B. At a constant dosage the strength of concrete containing type C PCMPW was improved up to 21%. At 1% addition of PCMPW by volume of mix in concrete, the strength improved up to 14% and in case of type C strength was significantly increased up to 38%. Beyond 1% of dosage of PCMPW, the strength continued to improve but at a relatively less significant rise. At 1.5% dosage the strength improved up to 11% for types A and B and up to 20% for type C. At full dosage of 2% the strength showed improvement in the range of 5%–14% including all three types of PCMPW.

4.3.10 PILOT STUDIES CONCRETE CONTAINING INDUSTRIAL WASTE AND PLASTIC WASTES

Preparation of the concrete from the industrial wastes, namely fly ash, silica fume, furnace slag along with the alkaline activator, is referred as the geopolymer-based concrete. The concrete so prepared requires significant investigations at the trial levels or as the pilot studies because it consists of different variables for chemical compositions, their proportions in the mixtures, and the treatment of the curing through which the material attains maximum strength known as the polymerization process during the curing with oven or at the ambient temperature. It is to be noted that changing the chemical composition of the solutions and the content of any of the basic pozzolanic constituent changes the properties of the mix to a significant extent.

4.3.11 TRIAL CONCRETE MIXES

In the beginning, numerous trial mixtures of geopolymer concrete were manufactured, and test specimens in the form of 150 mm × 150 mm × 150 mm cubes were made. Initially, the mixing was attempted in an electrical tilting drum mixture machine. The main objectives of the trial concrete mixes were:

- To familiarize yourself with the making of fly ash-based geopolymer concrete
- To understand the effect of the sequence of adding the alkaline liquid and metallized polymer plastic to the solid's constituents in the mixture
- To observe the behavior of the fresh fly ash-based geopolymer concrete and its workability by slump test and compaction factor test
- To develop the process of mixing

The object of trial mixes is to find the proportions in which the geopolymer concrete materials such as fly ash, alkaline liquid, extra water, fine aggregate, and coarse aggregate should be combined to provide the specific strength, workability, and durability with and without the metallized polymer plastic. The proportioning of ingredients of concrete is governed by the required performance of concrete in the plastic and the hardened stages. The objective of trial concrete mixes is to arrive at the economical and practical combination of different ingredients to produce geopolymer concrete using metallized polymer plastic waste that will satisfy the performance requirements under specific conditions of use. An integral part of concrete mix proportioning is the preparation of trial mixes and effect adjustment to such trial mixes to strike a balance between the requirement of workability, strength, and durability. Concrete must be of satisfactory quality in both stages, namely fresh stage and hardened stage. This mission is best practiced by trial mixes. During trial mixes different types of fly ash, sodium silicate solution as alkaline activator was used in geopolymer concrete, to check the feasibility of material used in the concrete mix. In this pilot study fly ash obtained from Gujarat thermal power station was used as source material and sodium hydroxide, chemicals, and sodium silicate from the locally available market.

In this study, we cast five cubes and two cylinders per mix. And we performed workability test on fresh concrete and compression test and split tensile strength

TABLE 4.18
Mix Proportion of Geopolymer Concrete

Trial Mix No.	Fly Ash (kg)	NaOH Solution (NaOH Solid+Water) (kg)	Na_2SiO_3 Solution (kg)	Coarse Aggregate (kg) 20 mm	10 mm	Fine Aggregate (kg)	Plastic (kg)	Extra Water (kg)
1	368	22.30+69.69	92	443.52	850	554	0	0
2	368	22.30+69.69	92	443.52	850	554	7.06	36.8
3	368	22.30+69.69	92	443.52	850	554	14.12	36.8
4	368	22.30+69.69	92	443.52	850	554	21.28	36.8
5	368	33.45+58.55	92	443.52	850	554	0	55.2
6	368	33.45+58.55	92	443.52	850	554	7.06	55.2

on hardened concrete. In our study we were only varying parameters such as sodium silicate to sodium hydroxide ratio, type of sodium silicate, molar intensity of sodium hydroxide, curing type, curing time etc. We used constant parameters such as sodium silicate to sodium hydroxide ratio and added extra water and finally measured the compressive strength of geopolymer concrete for every mix. The concrete mix design was carried out considering the following density of geopolymer concrete similar to that of OPC concrete, i.e., 2,400 kg/m^3. A generalized guideline to achieve desired strength of heat-cured low-calcium fly ash geopolymer concrete has been suggested in the literature (Hardjito et al. 2004; Rangan 2008; Sumajouw 2007). These guidelines were followed in the present work and various trial mixes were cast. Coarse aggregate and fine aggregate together were taken as 77% of entire mixture by mass (can be taken in the range of 75%–80%). While fine aggregate was taken as 30% of the total aggregate. Coarse aggregate of 20 mm size was taken for 24% of the total aggregate and 10 mm size was taken for 46% of the total aggregate. The details of material quantity required for 1 m^3 are given in Table 4.18.

In this pilot study we cast all concrete mixes using the same mix proportion for every mix change in molar content of NaOH, extra water, and curing type. The sodium silicate to sodium hydroxide ratio was 1. A total of six mixes were cast in the same ratio but in every mix the percentage of metallized polymer plastic waste and extra water was changed. Extra water of 0%, 10%, and 15% of fly ash by mass was added. Super plasticizer, 0.5% of fly ash by mass, was added in all trial mix.

4.3.12 MIXING

The sodium hydroxide (NaOH) flakes were dissolved in distilled water to make the sodium hydroxide solution. The NaOH solution was kept at 8 and 12 M for the trial mix. The sodium silicate solution and the sodium hydroxide solution were mixed 1 day prior to use to prepare the alkaline liquid. On the day of casting of the specimens, the super plasticizer was mixed with extra water to prepare the liquid component of the mixture. The fly ash and the aggregates were first mixed together in the electrical tilting drum mixer machine for about 3 minutes as shown in Figure 4.20.

FIGURE 4.20   Freshly mixed geopolymer concrete constituents.

The metallized polymer plastic flakes were then added (Figure 4.21), and the dry mixture is rotated for another 2 minutes. Alkaline solution along with super plasticizer and extra water was added to dry mix and mixing was done for about 2–3 minutes to achieve workable concrete mix. The test arrangements for compressive strength and splitting tensile strength are shown in Figure 4.22.

It was found that the amount of water in the mixture played an important role in the behavior of fresh concrete. When the mixing time was long, mixtures with high

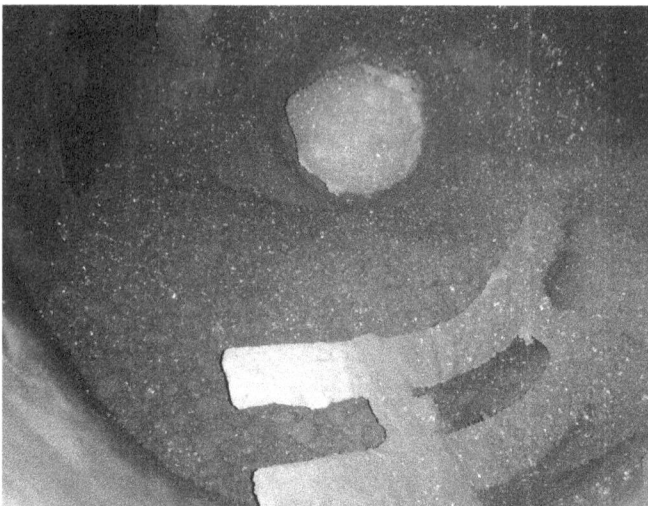

FIGURE 4.21   Metallized polymer plastic waste added in concrete.

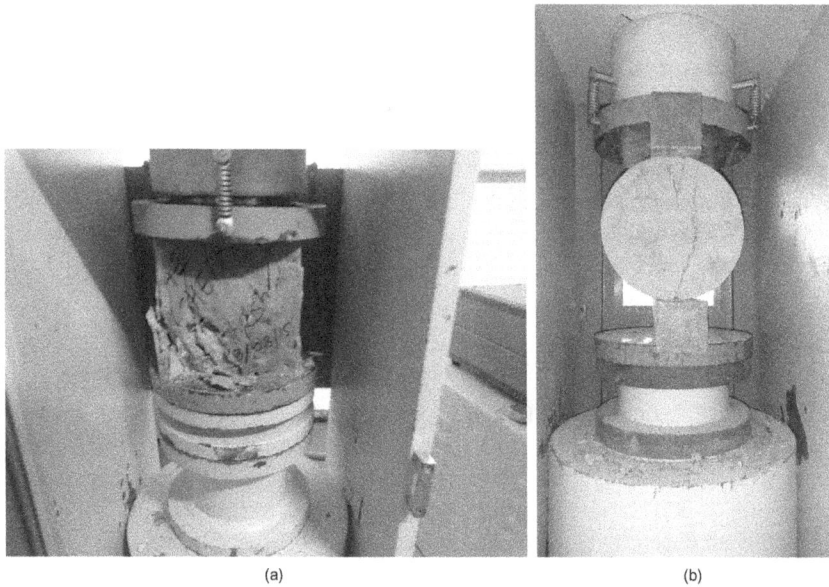

(a)   (b)

FIGURE 4.22   (a) Compression test. (b) Split tensile test.

water content bled and segregation of aggregates and the paste occurred. This phenomenon was usually followed by low compressive strength of hardened concrete. In this pilot study we observed that the specimens had problems during demolding. Concrete was stuck in the inner face of the specimens. Therefore, we stuck white cello tape on the inner face of all the molds, which was found to be in good condition even after hot air oven curing. From the pilot study, it was decided to adopt the following standard process of mixing in all further studies. Mix sodium hydroxide solution and sodium silicate solution together 1 day prior to casting of geopolymer concrete.

Mix all dry materials in the electrical tilting drum mixer machine for about 3 minutes. Add the polymer plastic and again dry mix it for a further 2 minutes. Add the liquid component of the mixture at the end of dry mixing and continue the wet mixing for another 2–3 minutes to get workable geopolymer concrete.

4.3.13 CURING

Preliminary pilot study tests revealed that fly ash-based geopolymer concrete did not harden immediately at room temperature (ambient curing). When the room temperature was less than 30°C, the hardening did not occur for at least 24 hours. Also, the handling time is a more appropriate parameter (rather than setting time used in the case of OPC concrete) for fly ash-based geopolymer concrete. Geopolymer concrete specimens should be wrapped during curing at elevated temperatures in a dry environment (in the oven) to prevent excessive evaporation. In trial mix curing temperature was 100°C at oven for 24 hours.

4.3.14 RESULTS

4.3.14.1 Compressive and Split Tensile Strength Test

Strength of concrete is measured in the hardened state. Strength of concrete is measured in a number of ways, such as strength in compression, in tension, in shear or in flexural, in abrasion, and in impact. In the pilot study, tests were performed to confirm required compressive strength and spilt tensile strength of fly ash-based geopolymer concrete after 28 and 7 days of ambient curing and hot air oven curing, respectively.

4.3.15 PILOT STUDY ON AMBIENT CURED GEOPOLYMER CONCRETE

Preliminary pilot study tests were carried out by varying the concentration of sodium hydroxide 3 M, 4 M and 5 M and 12 M. Alkaline liquid to fly ash ratio was taken as 0.5. Ratio of the activator chemical ($Na_2SiO_3/NaOH$) was taken as 2, and types of sodium silicate (SiO_2/Na_2O ratio) ratio was 3. Extra water: 10%, super plasticizer was 0.5% of fly ash. After casting, the test specimens were left open to air dry for curing at room temperature of 30°C (ambient curing) for 28 days. The average compressive strength of geopolymer concrete cube (with low concentration of sodium hydroxide 3 M–5 M) at 28 days was observed in the range of 2–4 MPa, which is not satisfactory. The average compressive strength of geopolymer concrete cube (with concentration of sodium hydroxide 12 M) at 28 days was observed to be 15.11 MPa, which is also not satisfactory.

The average tensile strength of geopolymer concrete cylinder was also much less as compared to that of conventional concrete. Another pilot study test was carried out by varying the concentration of sodium hydroxide, alkaline liquid to fly ash ratio was taken as 0.5, ratio of the activator chemical ($Na_2SiO_3/NaOH$) was taken 1, and types of sodium silicate (SiO_2/Na_2O): 2.25, varying the extra water, and super plasticizer: 0.5% of fly ash. Various parameters taken for the study and strength results obtained are given in Table 4.19.

TABLE 4.19
Testing Results of Trial Mixes

Batch Parameters	Batch B1	Batch B2	Batch B3	Batch B4	Batch B5	Batch B6	Batch B7
Liq./FA	0.5	0.5	0.5	0.5	0.5	0.5	0.5
$Na_2SiO_3/NaOH$	1	1	1	1	1	1	1
SiO_2/NaO_2	2.25	2.25	2.25	2.25	2.25	2.25	2.25
Molarity of NaOH	8	8	8	8	12	12	12
Metallized plastic	0%	0.5%	1%	1.5%	0%	0%	0.5%
Super plasticizer	0.5%	0.5%	0.5%	0.5%	0.5%	0.5%	0.5%
Extra water	0%	10%	10%	10%	10%	15%	15%
Curing temp.	100°C	100°C	100°C	100°C	100°C	100°C	100°C
Curing period in hours	24	24	24	24	24	24	24
Age of concrete	7 days	7 days	7 days	7 days	7 days	7 days	7 days
Density in kg/m³	2,461	2,318	2,313	2,263	2,310	2,267	2,205
Av. comp. strength in MPa	34.22	20.74	18.37	12.44	15.8	13.33	9.92
Split tensile strength in MPa	2.26	1.27	1.13	0.707		0.60	0.81

4.3.16 OBSERVATIONS

The average density of fly ash-based geopolymer concrete was observed to be similar to that of ordinary Portland cement concrete (2,400 kg/m^3).

The compressive strength of ambient cured geopolymer concrete was observed very less as compared to that of cured cement at hot air oven.

The addition of extra water improved the workability characteristics of geopolymer concrete mixtures but significantly decreased the compressive strength.

Addition of metallized plastic waste up to 1.5% by volume of concrete did not significantly affect the density and compressive strength of concrete.

Higher molar concentration of NaOH in solution results in a higher compressive strength. The compressive strength of hot cured concrete does not increase substantially after 7 days.

From the observation obtained during the pilot study, the geopolymer concrete using metallized polymer plastic waste up to 1.5% by volume of concrete did not significantly affect the strength of geopolymer concrete and hence it was decided to use polymer plastic waste as a constituent of geopolymer concrete and further study was carried out.

4.4 SIGNIFICANCE OF ADDITION OF WASTE IN CONVENTIONAL CONCRETE

The wastes in any form namely as constituents or fillers when added into the conventional mixtures impose alteration in the basic properties. However, to avail the advantages of waste utilization in a larger quantity, although mixing of wastes in concrete may sound good, care should be taken that any such addition must not adversely influence the basic properties of concrete and especially for the intended applications. The concept of addition of waste in concrete should be seen with respect to the following parameters or attributes to avail the significance in a broader means.

Mitigation of the use of natural materials in construction

Obtaining eco-efficient material by partial to full replacement of the natural materials

4.4.1 ENVIRONMENTAL SAFETY AND PROTECTION

Use of hazardous wastes in concrete in a quantified manner

Protection of the depletion of the natural material and the ecosystem

Value addition to the existing material namely concrete for improved life cycle and sustained use for extended service life

Reduction of the carbon footprints and emission of greenhouse gases otherwise produced in the manufacturing and processing of conventional concrete materials and mixtures

As discussed above, the primary concern of the use of waste in concrete is to mitigate the use of the conventional materials and utilize the abundant wastes as a solution to the hazardous impacts on the environment. There may be an additional advantage of the same in achieving the economy as well. Therefore, the development of concrete composites from waste should be explored more in detail with different materials as discussed in the next chapters of the book.

4.5 POTENTIAL OF MODIFIED CONCRETE
FOR DETAILED EVALUATION

The area of use of wastes in concrete holds a great potential for research, development, commercialization, manufacturing, and environmental safety. There are wastes, namely fly ash, silica fumes, used foundry sand, and plastics that may be replaced with the conventional material in any feasible fraction from 10% to 100% in a suitable mixture and by feasible means. Therefore, the topic should be studied further with global interest by all the academicians, researchers, scholars, professionals, and students.

REFERENCES

[1] Bhogayata, Ankur C., and Narendra K. Arora. "Fresh and strength properties of concrete reinforced with metalized plastic waste fibers." Construction and Building Materials 146 (2017): 455–463.

[2] Bhogayata, Ankur, and N. K. Arora. "Feasibility study on usage of metalized plastic waste in concrete." Contemporary Issues in Geoenvironmental Engineering: Proceedings of the 1st GeoMEast International Congress and Exhibition, Egypt 2017 on Sustainable Civil Infrastructures 1. Springer International Publishing, 2018.

[3] Bhogayata, Ankur C., and Narendra K. Arora. "Impact strength, permeability and chemical resistance of concrete reinforced with metalized plastic waste fibers." Construction and Building Materials 161 (2018): 254–266.

[4] Bhogayata, Ankur C. "Concrete reinforced with metalized plastic waste fibers." Use of Recycled Plastics in Eco-efficient Concrete. Woodhead Publishing, 2019. 349–367.

5 Material Properties of Concrete Containing Plastic Wastes

5.1 MIX DESIGN OF THE CONCRETE MODIFIED WITH PLASTIC WASTES

Three mixes were prepared with varying water to cement ratio of 0.45, 0.55, and 0.65. The mix design conformed with IS: 10262-2021. The mixed proportions shown in Tables 5.1 and 5.2 provide the details regarding the mix design and batches prepared for the experimental study.

It is to be noted that the PCMPW fibers are of average sizes and require special care during the shredding process for the screen blade or cutter blade. The width is also measured as the average dimension depending on the shredder blade size.

TABLE 5.1
Mix Proportions for 1 m³ Concrete

Mixture	Cement	Aggregates 20 mm	Aggregates 10 mm	Sand	Water	W/C ratio
	kg	kg	kg	kg	kg	
1	420	670	440	640	185	0.45
2	355	675	450	695	194	0.55
3	305	660	445	715	198	0.65

TABLE 5.2
Batch Designations and PCMPW Details

Water to Cement Ratio	PCMPW Fiber Type	Batch designation	PCMPW fiber fraction
0.45	A	B1 to B5	0% to 2%
	(1 mm)	B16 to B20	0% to 2%
		B31 to B35	0% to 2%
0.55	B	B6 to B10	0% to 2%
	(5 mm)	B21 to B25	0% to 2%
		B36 to B40	0% to 2%
0.65	C	B11 to B15	0% to 2%
	(20 mm)	B26 to B30	0% to 2%
		B41 to B45	0% to 2%

DOI: 10.1201/9781032621340-5

5.2 METHODS OF PREPARING CONCRETE WITH PLASTIC WASTES

Material performance was assessed for alteration in standard property values due to the inclusion of PCMPW in conventional concrete. Tests for workability, strength, durability, and structural response were performed in laboratories. Standards relevant to the type of tests were referred to including IS codes and ASTM. Fresh behavior was assessed based on slump and compaction factor test values. Hardened concrete molded into cube, cylinder, disk, and beam specimens was tested to obtain compressive strength, splitting tensile strength, impact strength, and pull-off strength. Impact resistance was examined by drop weight method. Pull-off strength or the bond strength was determined with standard guidelines of ASTM C 1583-04. Resistance to acid, sulfate, and chloride ingress in concrete containing varying fractions and sizes of PCMPW was assessed in accordance with standard guidelines of ASTM codes, respectively.

Permeability tests were conducted on disk specimens developed in the laboratory. Disk specimens were submerged into water to study the rate of water absorption by concrete containing PCMPW flakes.

Experimental values obtained for compressive strength and splitting tensile strength were compared to obtain analytical interrelationship of the quantities. Moreover, the stress and corresponding strain values obtained by the tests on cylinder specimens were used to obtain their interrelationship. Reinforced beam specimens were subjected to flexure to obtain the moment-resisting capacity and alterations due to the PCMPW inclusion. Analytical and experimental results were compared for obtaining the response of beams in flexure as structural application of proposed material. Table 5.3 shows details of specimens prepared for testing.

TABLE 5.3
Details of Test Specimens

Test	Specimen	Dimensions (mm)			Total Numbers
		L	B	H/D	
Compressive strength	Cube	150	150	150	171
Pull-off strength	Cube	150	150	150	45
Acid resistance	Cube	150	150	150	45
Sulfate resistance	Cube	150	150	150	45
Splitting tensile strength	Cylinder	–	150	300	135
Stress–strain relationship	Cylinder	–	150	300	12
Impact strength	Disk	–	150	60	45
RCPT	Disk	–	100	60	45
Oxygen permeability	Disk	–	100	60	45
Water sorptivity	Disk	–	100	60	45
Flexural response	Beam	1,500	150	200	12

TABLE 5.4
Standard Reference Values for Slump

Slump Value in mm	Category	Application or Use
100–150	High	Pumping, piling, trench-fill
50–100	Medium	Primary structural elements
25–75	Low	Mass concrete, pavements
Less than 25	Low	Shallow foundations

5.3 FRESH PROPERTIES

The workability of fresh mix of concrete containing PCMPW flakes in varying proportions was measured by slump tests. The mixing process is shown in Figure 4.4. Slump behavior indicates stiffness of fresh mix to work with. Factors like size and shape of aggregates and water to cement ratio primarily affect the slump behavior. Slumps are measured as high, medium, and low. Slump values were obtained for every mix before further usage in testing of properties. Slump is a variation in height of the fresh concrete held in the set position by a conical mold with reference to the standard values. Table 5.4 shows standard slump values and usage of concrete.

5.3.1 SLUMP TEST

Slump is maintained according to the need of application. However, slump must be controlled to keep minimum workability of fresh mix. Concrete reinforced with PCMPW exhibited different values for slump for varying size and fraction. The water to cement ratio remained almost standard as the low values reduced the slump and high values increased slump.

5.3.2 OBSERVATORY NOTES

Slump of concrete was found sensitive to inclusion of PCMPW. Slump response varied with the test parameters. While preparing the concrete for slump test, it was observed that the flakes tend to float on the surface of mix and showed no adherence with constituents at the first stage, hence more time was required for mixing the constituents. With time and number of revolutions, the flakes started getting mixed with the constituents. The mix became stiffer on increased contents of PCMPW. Effects of type of PCMPW on slump were observed carefully. Addition of PCMPW type B showed more reduction in slump compared to other types. Inclusion of higher volume fractions of flakes resulted in non-uniform mixes. PCMPW increased balling effects resulting in segregation of aggregates from the matrix. However, small-sized flakes remained ineffective on mixing response of constituents at dosage up to 1%. A similar response was noticed in the case of type C PCMPW (Figures 5.1 and 5.2).

Slump test results are shown in Figure 5.3. The figure shows variation of slump values with respect to PCMPW fractions for three different water cement ratios.

FIGURE 5.1   Addition of the PCMPW fibers into the concrete mixture.

FIGURE 5.2   Concrete slump measurement containing PCMPW fibers.

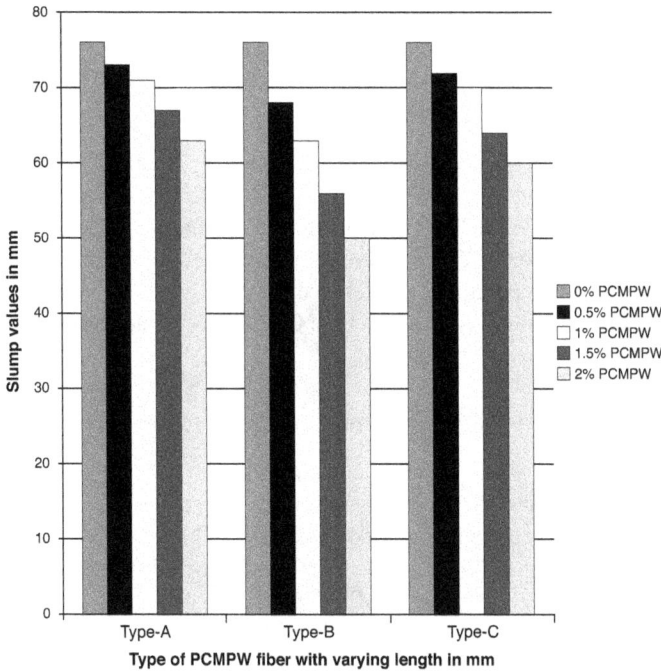

FIGURE 5.3   Effect of PCMPW type on slump of concrete.

In all mixes, inclusion of PCMPW in concrete reduced the slump, which is visible from the trend. Slump reduction indicated that addition of PCMPW makes concrete stiff and less viscous. The viscosity of a mix improves spreading of concrete in formworks and reduces air voids in a mix. Some concrete may not be workable if slump continues to reduce for varying fractions of PCMPW. Therefore, it was required to obtain the optimum PCMPW dosage and type.

Results showed that water to cement ratio did not affect the trend of reduction of slump for a given PCMPW attributes, namely fraction and type. According to the test results, the most significant parameter for slump response was the fraction of PCMPW. Moreover, the trends of slump reduction were similar for concretes containing type A and type C PCMPW, and type B showed larger reduction of slump relatively.

The effect of addition of 0.5% of type A and type C PCMPW reduced slump up to 4% and continued to reduce up to 7%, 12%, 14%, and 21% for every 0.5% increment in fraction of PCMPW till 2% by volume of concrete. Type B reduced slump up to 34% of addition at 2% addition by volume of concrete. Slump results exhibited relatively less reduction for type A and type C PCMPW up to 1% addition by volume of concrete.

PCMPW of type B was found responsible for increased loss of viscosity by restricting homogeneous mixing of constituents. Larger size of flakes as type B plastic increased segregation of aggregates from cement paste and increased stiffness of a mix. The results showed that the optimum dosage of PCMPW should be 1% with type A or type C PCMPW to be used in concrete to maintain the workability.

5.3.3 EFFECT OF PCMPW TYPE

Figure 5.3 shows the variation in slump values with type of PCMPW like type A, type B, and type C for all three mixes. Type B PCMPW reduced the slump more than the other two types of PCMPW for the given fraction and water cement ratio values.

At a constant fraction of 2% of PCMPW by volume of concrete mix, slump was reduced up to 34% compared to the reference concrete due to the presence of type B plastics. On the other hand, type A and type C PCMPW reduced slump up to 21% at the dosage of 2%.

5.3.4 EFFECT OF WATER CEMENT RATIO

Figure 5.4 shows the effect of water cement ratio on slump of concrete containing PCMPW. The water cement ratio variation affected the slump significantly. Increased water cement ratio increased the slump. This response was seen in all three mixes containing PCMPW. However, the addition of PCMPW contributed to the reduction of slump for a given PCMPW type.

The slump of concrete reduced up to 38%, 34%, and 21% for the mixes prepared with 0.45, 0.55, and 0.65 water cement ratio. For a given fraction of PCMPW, the maximum slump was reduced in case of type B.

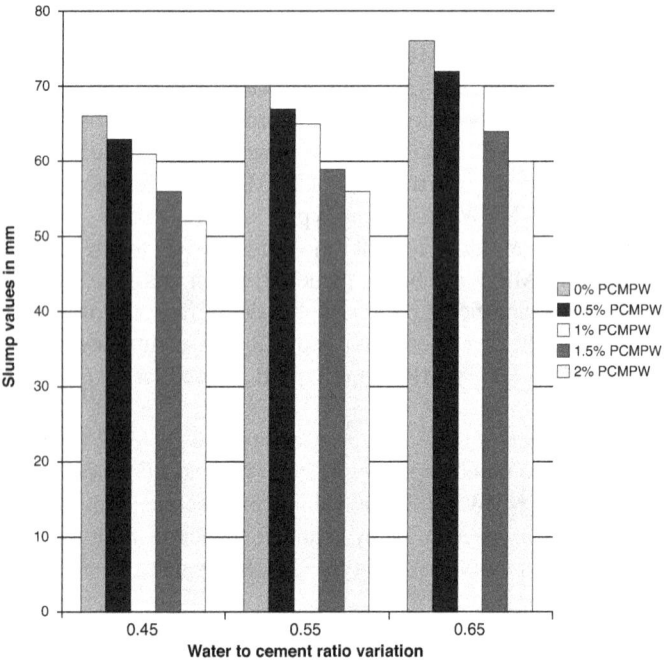

FIGURE 5.4   Effect of water to cement ratio on slump values of modified concrete.

5.3.5 COMPACTING FACTOR TEST

Compaction factor test was required when the concrete is relatively stiff and shows less slump (IS: 1199-1959). The test is sensitive for workability measurement and response as it accounts for the effects of initial hydration process of cement in concrete. Freshly mixed concrete should be promptly used for the test to obtain accuracy in results. Like slump values, Table 5.5 shows ranges of compacting factors according to their usage. Figure 5.5 shows the test apparatus and concrete modified with plastic fibers.

TABLE 5.5
Standard Reference Values of Compacting Factor

Compacting Factor Test	Category	Application or Use
0.95	High	Pumping, piling, trench-fill
0.92	Medium	Primary structural elements
0.85	Low	Mass concrete, pavements
0.78	Very low	Shallow foundations

FIGURE 5.5 Compacting factor test on concrete fresh mixture.

5.3.6 OBSERVATORY NOTES

Conforming to the requirements of IS: 1199-1959 codes, standard apparatus was used for obtaining compacting factor values of each mix. The time of shutter release was strictly followed not more than 2 minutes between the two consecutive hoppers. It was observed that concrete containing type B flakes showed less free flow while allowing the drop of concrete from the upper hopper to the lower hopper. Figure 5.5 shows a compacting factor test being carried out.

Compaction factor test results are shown in Figure 5.6. It shows a gentle reduction in values with increased PCMPW fractions in the range of 0%–2% by volume of the mix. The results represented in graphical format reveal very little variation in compaction factor values by varying test parameters (Figure 5.7).

Concrete containing type A and type C of PCMPW showed almost similar and gradual trend of reduction, while concrete containing type B differed in response showing steep reduction trend lines. This scenario was observed as common for all the three types of concrete. It was observed that concrete prepared with 0.45 as water to cement ratio showed minimum reduction in values of compacting factor among all three mixes (Figure 5.8).

Compaction factor reduced up to 5% at 0.5% addition of PCMPW by volume of concrete. The values reduced up to 10% at 1% addition of PCMPW and continued to reduce up to 20% at full dosage of PCMPW at 2% by volume of concrete. It was observed that maximum reduction of 22% of values was in case of PCMPW type B for all three types of concrete mixes.

PCMPW type A and type C reduced the values up to 15% as maximum reduction, while type B reduced values up to 22%. At dosage of 2% of PCMPW, types A and C showed minimum reduction of 11% specifically in case of concrete prepared with 0.45 as water to cement ratio.

Compaction factor test results reduced with increased PCMPW fractions. The presence of PCMPW obstructed the rearrangement of aggregates in the mix and expelling of entrapped air in the mix. Concrete mixes showed segregation of constituents at higher dosage of PCMPW and reduced degree of compaction. The free falling of concrete mass containing PCMPW between the hoppers was also affected with increased PCMPW fractions.

A careful observation of test results and experience of working with fresh concrete containing PCMPW during the compaction factor tests led to the observation that PCMPW can be added to the concrete with a dosage of 1% by volume of mix as the consistency remained same at this dosage in all mixes. The preferred dimension of PCMPW particles should be about 1 mm average as flake or 1 mm × 20 mm fiber form.

5.3.7 EFFECT OF PCMPW FRACTION

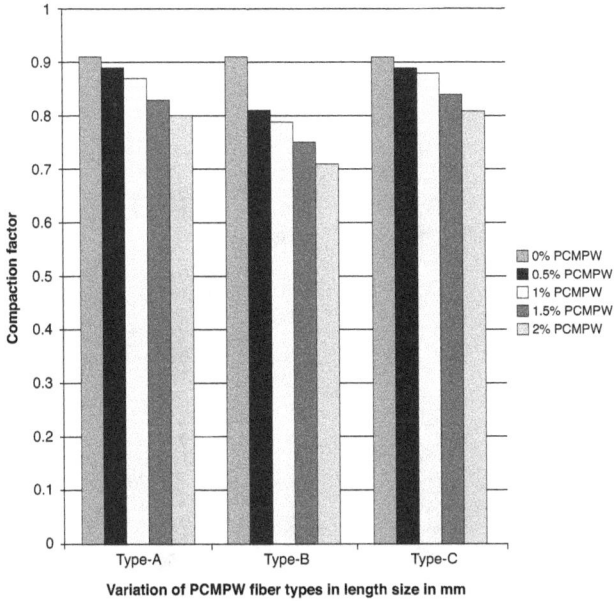

FIGURE 5.6 Effect of PCMPW fractions on compaction factor of concrete.

5.3.8 EFFECT OF PCMPW TYPE

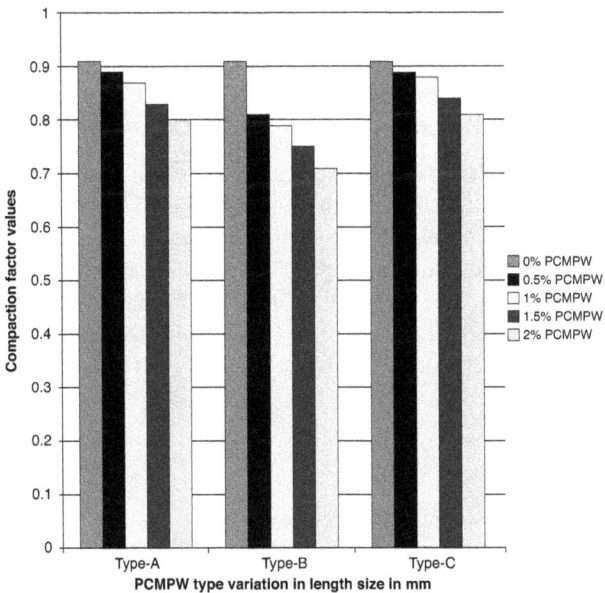

FIGURE 5.7 Effect of PCMPW type on compaction factor of concrete.

5.3.9   EFFECT OF WATER CEMENT RATIO

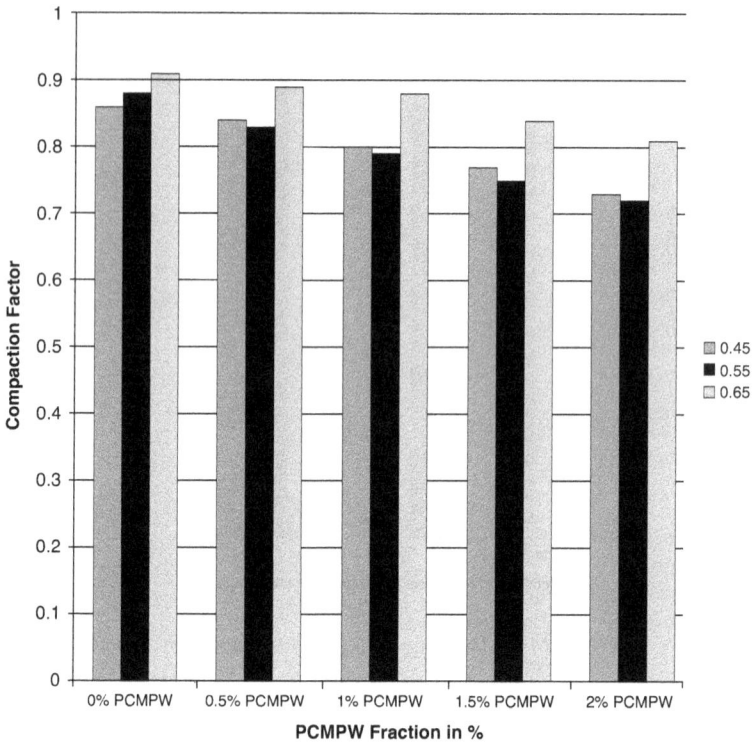

FIGURE 5.8   Effect of water cement ratio on compaction factor.

5.3.10   RELATIONSHIP BETWEEN COMPACTION FACTOR AND SLUMP

Slump and compaction factor values were used to establish a relationship between the workability parameters. Without distinguishing the PCMPW flake sizes or water cement ratio, the analytical relationship was established as shown in Figure 5.9.

It was found that compaction factor and slump values were in linear parabolic relationship analytically. The trend was found to be parabolic, and relationship was governed by the degree of compaction of the mixes. Such a relationship has been observed by other research during investigations on fiber-reinforced concrete.

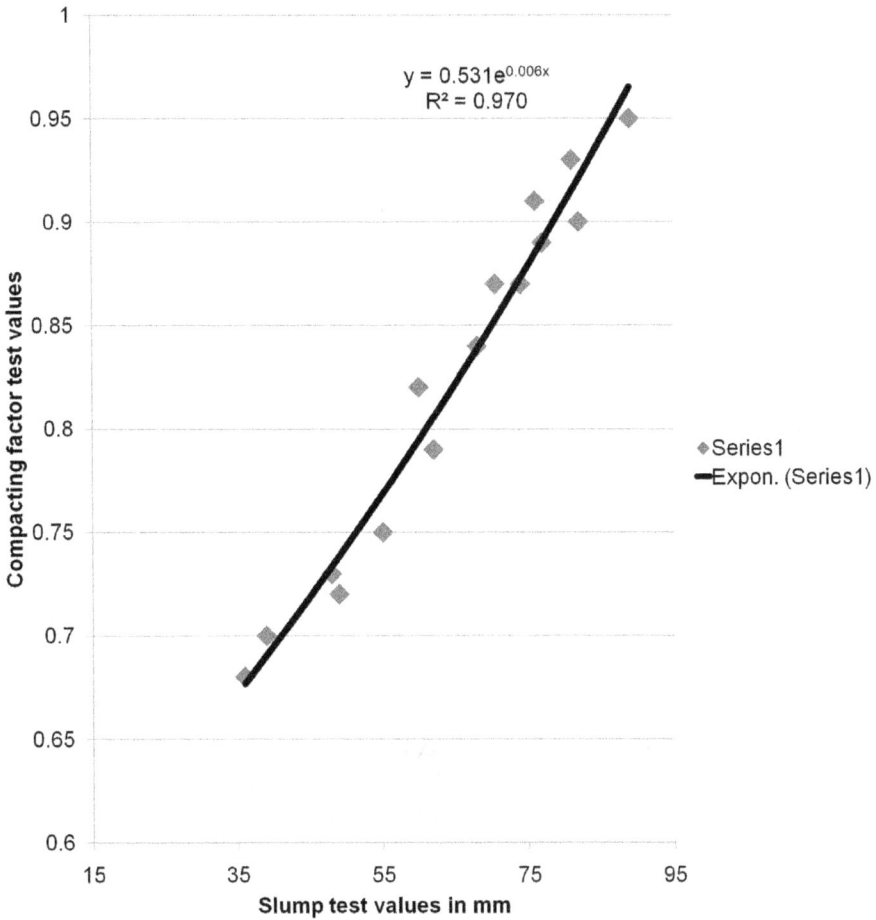

FIGURE 5.9   Relationship of compaction factor with slump values.

5.3.11   SIGNIFICANCE OF WORKABILITY TESTS

Workability properties were studied to receive the fresh state behavior of concrete to evaluate the effects of addition of PCMPW in concrete.

Test results demonstrated potential usage of PCMPW in concrete up to the fraction of 1% by volume of a mix. Workability properties namely slump and compaction factor reduced up to 23%.

The preferred types of PCMPW were flakes of 1 mm average size and in fiber form of 1 mm × 20 mm size.

Slump and compaction factor test results demonstrated that PCMPW could be used in the applications where stiff concrete mixes are desired as addition of PCMPW made the conventional concrete stiff.

The presence of PCMPW added voids to the mixes and increased the overall void volume and exhibited limitations in compaction capacity of concrete. This resulted in the reduction of compacting factor values with increased flake contents.

For structures like pavements, linings, and other similar mass concrete works where relatively stiff mixture is required, higher dosage of PCMPW in concrete can be practiced.

The use of concrete containing PCMPW can be effectively used for small scale nonstructural precast elements like fencing poles and modular walls.

5.4   MECHANICAL PROPERTIES

The mechanical properties are also regarded as strength properties. For concrete the strength properties are those that provide resistance against different loading conditions, namely compression, tension, flexure, and shear actions of the forces. Therefore, the ability of the material and matrix to resist these effects is necessary to evaluate the performance of hardened concrete under various loads. In this section the strength properties have been described in detail with results for the concrete modified with the PCMPW.

5.4.1   COMPRESSIVE STRENGTH

Cube and cylinder specimen conforming to the requirements of IS: 516-1999 was prepared to obtain strength of concrete in axial compression. A compression testing machine of 2,000 kN capacity was used to determine compressive strength of hardened concrete modified by PCMPW flakes. All molds were oiled and tightened against leakage. Specimens were water cured for 28 days at room temperature. An average value of three tests was recorded.

5.4.2   OBSERVATORY NOTES

While performing the tests, the reference concrete failed abruptly by showing instantaneous cracks. The response was different in the case of concrete containing varying fractions of PCMPW flakes. Regardless of the water to cement ratio and types of flakes, the concrete containing low fractions of PCMPW, i.e., 1% by volume of concrete, showed uniform micro cracks on exterior surfaces. At final failure also, the specimen remained bounded and demonstrated ductile failure including bleeding of concrete (Figure 5.10).

Compressive strength is referred to as one of the primary and most important properties of concrete. Compressive strength of concrete containing PCMPW was investigated for possible changes in the values due to the addition of PCMPW. Compressive strength was measured on cubes and cylinders.

Figure 5.11 shows variation in the compressive strength with the varying PCMPW fractions and types for all the three mixes. The trend lines indicated that the strength reduction scenario was nearly similar for all concrete mixes prepared with different water to cement ratio, namely 0.45, 0.55, and 0.65.

Concrete prepared with water to cement ratio of 0.45 showed minimum strength reduction at a constant fraction and type of PCMPW. The strength reduced as

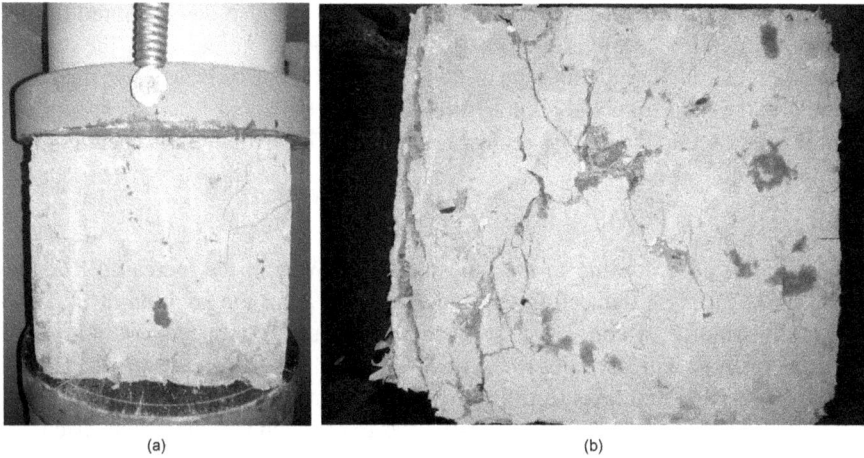

FIGURE 5.10   (a) Compressive strength test arrangement. (b) Tested specimen.

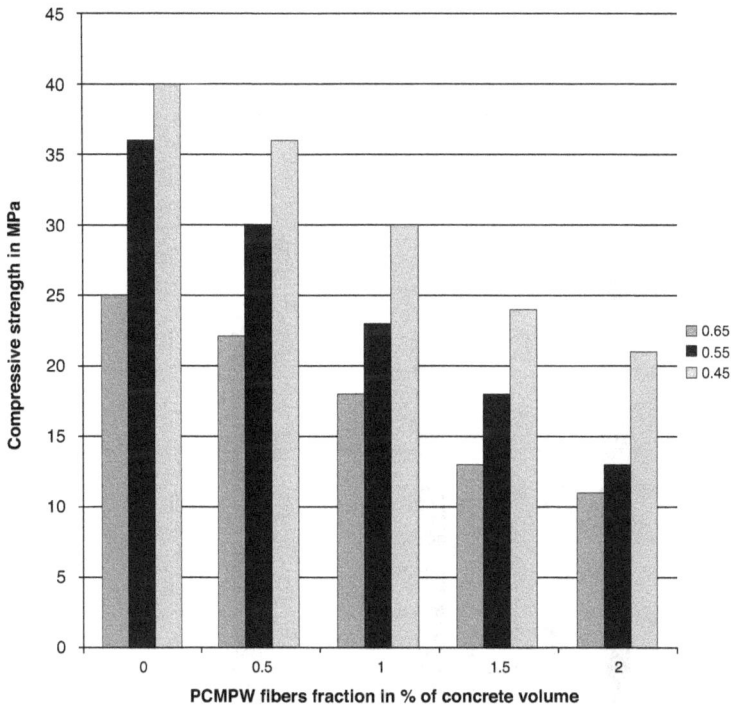

FIGURE 5.11   Effects of PCMPW fibers on compressive strength.

negligible as 3% to the maximum of 18% for the tests conducted on specimens prepared by mix 1 (w/c: 0.45). However, the strength reduced in the range of 5%–29% for the dosage range of PCMPW from 0.5% to 2% for the same mix in case of type B PCMPW.

Concrete prepared with water to cement ratio of 0.55 also demonstrated limited reduction of 3%–9% at lower dosage of 0.5%–1%. However, unlike mix 1, the strength reduction scenario varied with type of PCMPW and type C of PCMPW reduced the strength in the range of 6%–12% at a lower dosage of 0.5% and 1%, respectively, and showed larger reduction of 24% at final fraction of 2% by volume of concrete.

5.4.3 EFFECT OF PCMPW FRACTION

Strength reduced according to the types of PCMPW with the increased fractions of PCMPW. Strength reduced in the range of 3%–8% for the addition of PCMPW of 0.5% by volume of concrete. The strength was further reduced in the range of 9%–12% with the increased dosage of PCMPW as 1% by volume of the mix. For the increased PCMPW fractions from 1% to 2% by volume of concrete, the compressive strength reduced in the range of 18%–24%. It is to be noted that the results discussed are for the mixes consisting of PCMPW of type A and type C.

Compressive strength reduced significantly by addition of PCMPW type B relatively and reduced the strength in the range of 9% to a maximum as 44% for the PCMPW fraction range from 0.5% to 2% by volume of concrete. The strength decreased from 8% to 44% at 0.5% and 2% dosage of PCMPW, respectively (Figure 5.12).

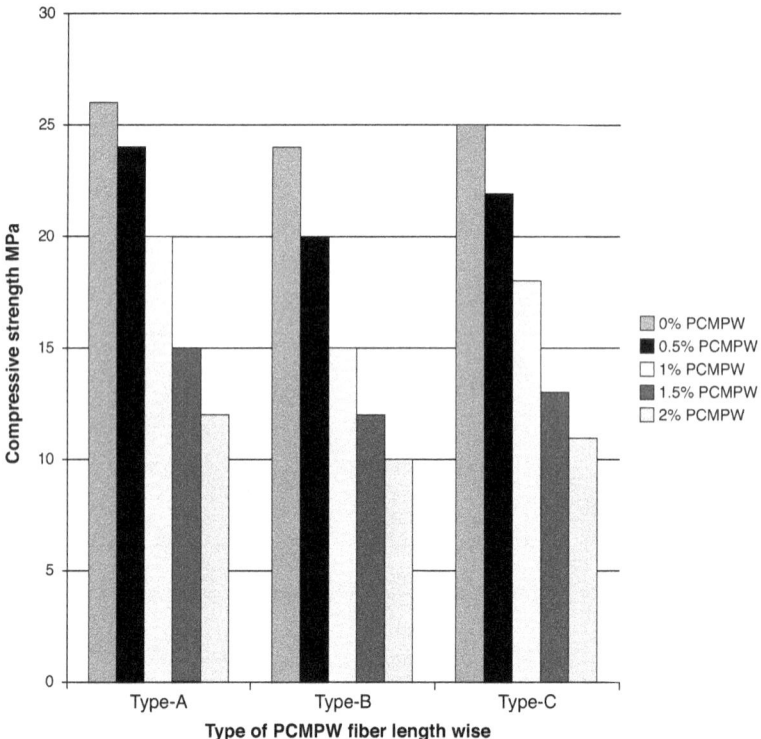

FIGURE 5.12   Effect of PCMPW type on compressive strength of concrete.

5.4.4 EFFECTS OF PCMPW TYPES

See Figure 5.12.

5.4.5 EFFECTS OF WATER CEMENT RATIO

PCMPW was shredded into three varying types, namely type A, B, and C, consisting of varying dimensions and forms as specified in the section on materials. Type A and type C of PCMPW showed very similar trends and range of strength reduction as is evident from the results of compressive strength shown in Figure 5.13. Type B of PCMPW reduced the strength relatively more than the other two types of PCMPW. Type A PCMPW contained relatively small-sized flakes of 1 mm average dimensions. Type C was in the form of fiber of 1×20 mm dimensions. These two types of PCMPW did not affect the strength significantly and for a low dosage of 0.5% and 1%, the reduction was within the upper limit of 12% only.

However, type B PCMPW in flake form of 5 mm average size showed more reduction in the strength at constant test parameters, namely water to cement ratio and fractions of PCMPW. Fore more discussions on this responses kindly refer to the work published by author [1]

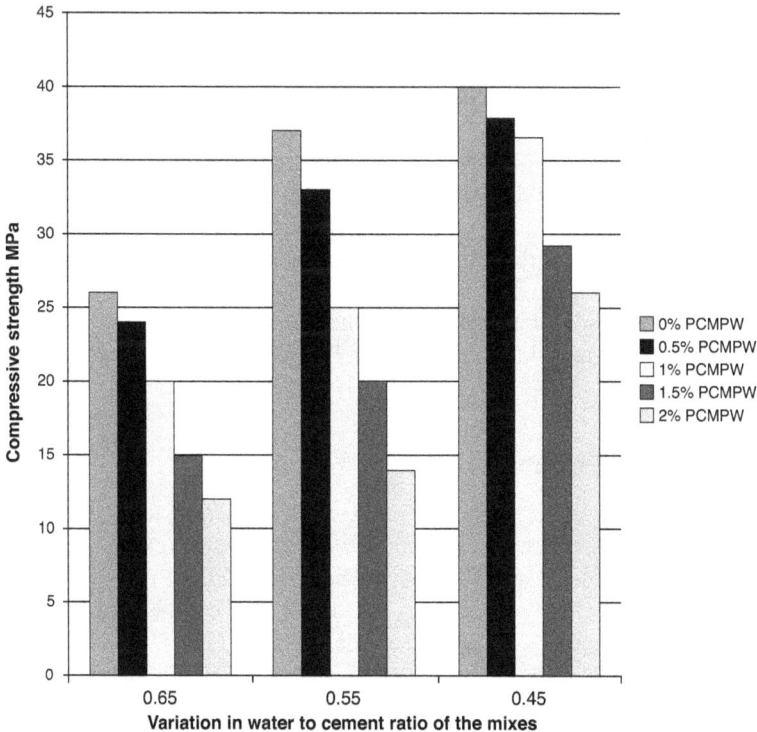

FIGURE 5.13 Effect of water cement ratio on compressive strength of concrete.

5.4.6 Significance of Addition of PCMPW on Compressive Strength

Addition of PCMPW in concrete reduced the strength. However, it was noticed that the reduction was insignificant for up to 1% of addition of flake-type PCMPW. It is important to identify the best dimensions and fractions of any non-conventional material to be added in concrete and the strength results provided important guidelines for such addition of PCMPW in concrete. It was noticed that type B flakes with large size of 5 mm or more were found unsuitable for usage.

5.4.7 Splitting Tensile Strength

Cylinders were tested for splitting tensile strength in accordance with IS: 5816-1999 for obtaining strength of concrete under indirect tension. Response of concrete from crack formation at the initial stage and final failure was studied. The test measures crack resisting capacity of concrete by indirect compression applied across the longitudinal axis of the specimen. Splitting tensile strength is expressed by fraction of compressive strength. Resistance of concrete to crack formation depends on compressive strength of concrete but at the transition zone around the aggregates and cement paste molecular adhesion plays a major role. Tensile failure of concrete is a resultant effect of failure of cement paste.

5.4.8 Observatory Notes

Crack patterns were carefully observed during the tests. Initial minor cracks across the diameter were gradually widened with increased loading in case of reference concrete. PCMPW affected cracking patterns especially in case of type A and type C, and was found capable to get a grip on hardened mass. Type C showed significant resistance against cracking and was found effective in reducing the crack propagation. Tensile resistance of concrete is a critical property of concrete. The addition of PCMPW was therefore studied for its effects on the tensile resistance of concrete mixes. Figure 5.14 shows changes in split tensile strength values of concrete due to the addition of PCMPW. The trend lines exhibited an increase in splitting tensile strength due to addition of PCMPW in concrete up to a specific fraction value.

5.4.9 Effects of PCMPW Fraction

Addition of PCMPW of 0.5% dosage improved the strength in a range of 10%–14% in case of type A and type B. At a constant dosage the strength of concrete containing type C PCMPW improved up to 21%. At 1% addition of PCMPW by volume of mix in concrete, the strength improved up to 14% and in case of type C strength was significantly increased up to 38%. Beyond 1% of dosage of PCMPW, the strength continued to improve but at relatively less significant rise. At 1.5% dosage the strength improved up to 11% for type A and type B and up to 20% for type C. At full dosage of 2% the strength showed improvement in the range of 5%–14% including all three types of PCMPW (Figure 5.15).

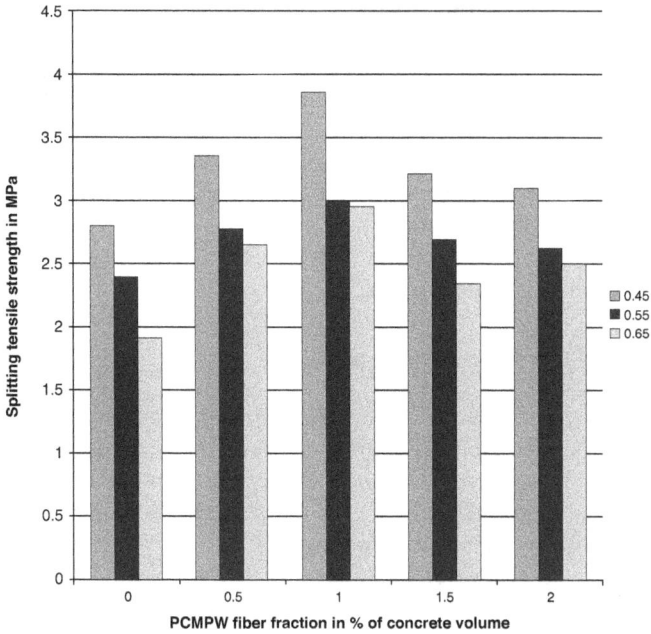

FIGURE 5.14 Effect of PCMPW fraction on splitting tensile strength.

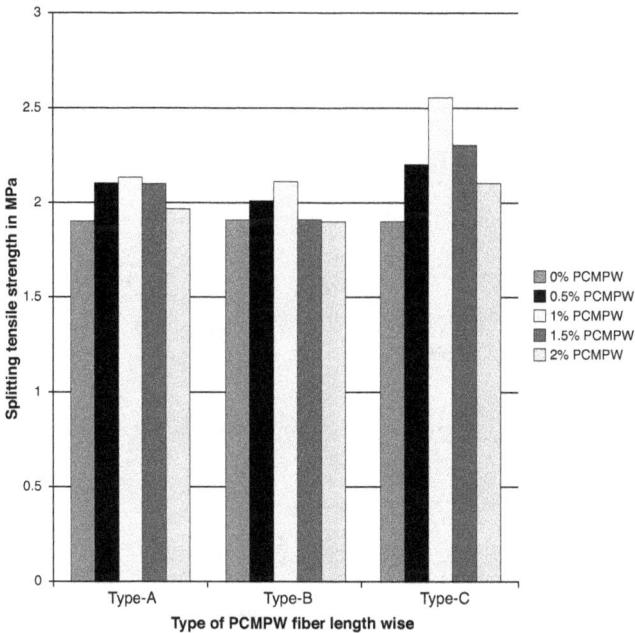

FIGURE 5.15 Effect of PCMPW type on splitting tensile strength.

5.4.10   EFFECTS OF PCMPW TYPES

Results revealed that type C PCMPW improved the strength most significantly up to 38% compared to the other types. The fiber form of PCMPW developed better bonding with the hardened mass of concrete. Cracks developed at the micro level were obstructed by type C PCMPW and the progress of micro cracks was restricted. At a constant type of PCMPW, the splitting tensile strength of concrete was improved compared to the reference concrete in all conditions, viz. varying water to cement ratio and varying PCMPW fractions. However, the strength improvement was not significant in the case of type B flakes.

5.4.11   EFFECTS OF WATER CEMENT RATIO

Split tensile strength has been referred to as the manifestation of strength in indirect tension and has been found to be dependent on the compressive strength of the concrete. It could be seen that the lower the water cement ratio, higher is the compressive strength and so is the splitting tensile strength. The mix prepared with 0.45 water cement ratio showed maximum strength for all sizes and fractions of PCMPW among all three mixes. However, inclusion of PCMPW improved the strength for all values of water to cement ratio under consideration (Figure 5.16).

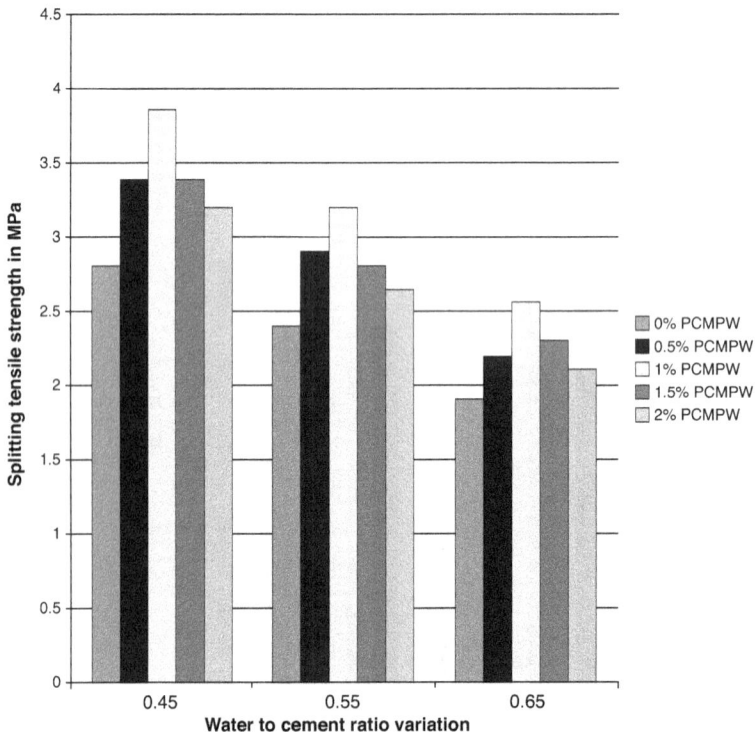

FIGURE 5.16   Effect of water cement ratio on splitting tensile strength.

5.4.12 PULL-OFF STRENGTH

The test determines the tensile strength or precisely the bond strength of concrete constituents in direct tension near the surface of adhesion with the adhesive applied on the surface. ASTM C-1583-04 was referred for the test. Cubes were used to carry out the test. The cubes were centrally drilled with a core of 50 mm diameter and 50 mm depth with a core drilling device. The core top was glued with an adhesive material and the machine capable of applying direct tension was fitted on the core with a metal capping. The setup was left for proper adhesion for 24 hours. With the provision of a rotating arm of apparatus aligned to the 60 mm height from the surface of cube, direct force was applied until the core separated from the cube and broke. The load applied and the resistance from the concrete were measured as the pull-off strength in N/mm^2 unit. The calculation was direct and involved the simple formulation as

$$\text{Pull off strength } (f_{\text{pull}}) = P/A$$

where P = applied direct load and A = cross-sectional area of the core.

Pull-off strength could be measured on the cylinder specimens also by developing the core at the center. The test utilizes the capacity of the concrete constituents to resist the direct tensile forces applied by means of a small manually operated machine. It is the measure of the internal adhesion and surface bonding of the constituents with the cement gel in the mix. The test was employed to check the effects of the presence of PCMPW flakes in the matrix of concrete and their intermolecular adhesion characteristics.

The test was performed on all batches of all three mixes and the results were received in the form of pull-off strength versus PCMPW flake contents and variation in the sizes of flakes. This test provided the information and response of the material against the direct application of the tensile force and the resistance governed by the molecular bonding of the constituents. Results were obtained and tabulated for varying contents of each batch and mix.

5.4.13 OBSERVATORY NOTES

Specimens containing type A and C flakes provided better bond and adhesion against the direct tension. More stress was generated to develop failure in the specimens containing PCMPW compared to the reference concrete. The core surfaces were employed with surface epoxy adhesives 24 hours before the tests to achieve proper test results. Intermolecular bonding of concrete constituents bears high importance during the combined action of loads to the structural members resulting in the concrete cracks and separation of hardened concrete. Propagation of micro cracks may result in failure particularly during persistent loadings. Determination of alterations in the adhesion capacity of concrete constituents containing PCMPW was important to be investigated. A pull-off strength test was conducted on cubes containing varying PCMPW types and fractions.

5.4.14   Effect of PCMPW Fraction

Pull-off strength increased in the range of 8%–14% at 0.5% fraction by volume for a constant PCMPW type. For the increased PCMPW dosage of 1%, the strength continued to increase up to 20%. Beyond 1% of PCMPW fraction, the strength showed reduction from 20% to 8%. For maximum dosage of 2% fraction of PCMPW, the strength reduced from 8% to 5%. Interestingly, the strength of concrete containing maximum fraction of PCMPW exhibited improved pull-off resistance compared to the reference concrete.

5.4.15   Effect of PCMPW Type

PCMPW consisting of type C exhibited maximum resistance. The result shows a trend of increased pull-off resistance up to 1% addition of PCMPW. Beyond 1% dosage the strength reduced, though it did not reduce more than the reference concrete response. Type B PCMPW improved the strength up to 9% at the dosage of 1%. Beyond 1%, the strength reduced significantly to 3%.

5.4.16   Effect of Water Cement Ratio

Pull-off strength exhibits the resistance offered by concrete to a direct tensile force. Strength of concrete is primarily governed by the cementing capacity of the paste to hold the constituents in the hardened state. Concrete prepared with lower water to cement ratio like 0.45 demonstrated excellent improvement against pull-off actions in the presence of PCMPW. A minimum resistance was offered by concrete prepared with 0.65 water cement ratio.

5.5   DURABILITY PROPERTIES

Chemical ingress results in degradation of strength of concrete. Concrete surface in contact with air, water, or soil faces chemical attacks in more than one way. Cement paste is largely attacked by acids and bond strength is reduced. Sulfate attacks on cement paste also and internal volumetric expansion takes place.

The role of PCMPW in acid and sulfate resistance was investigated by curing the concrete cubes containing varying PCMPW flakes in solutions prepared by acid and sulfate inclusion. Cubes of size of 150 mm × 150 mm × 150 mm were immersed in the acid and sulfate solution in the curing tank for 60 days. A large curing tank was prepared on temporary bases in labs for complete immersion of the cube specimens in these solutions.

5.5.1 Acid Resistance

The acid attack test was performed in accordance with IS: 1111-1987. Cubes were placed in curing tank filled with water solution containing 5% concentric sulfuric acid and 5% hydrochloric acid added to water. The curing continued up to 60 days at room temperature. The weight difference after and before curing was obtained after curing period by cleaning the surface of cubes. The weight reduction was referred to as acid attack on the surface of concrete.

5.5.2 Observatory Notes

The source of acid ingress was found to be dependent on void structure of the concrete. Test results were obtained in two parts. First was by weight loss of specimens and second was by measuring the compressive strength of cubes after cleaning the surface. Results were studied for possible effects of the presence of PCMPW flakes in concrete and the relevant effects on acid resistance.

5.5.3 Sulfate Resistance

The sulfate attack test is important for the durability of concrete. Sulfates can get in contact with concrete by many means like from salinity of the air, water, or soil. The evidence of sulfate attack on concrete has been observed as spalling of edges and surface and development of whitish form like layers on surface like effloresce. The hardened mass of concrete experiences the expansion of volume due to the chemical reaction of sulfates with the cement paste and formation of gypsum and calcium sulfo-aluminates. The most problematic sulfate has been magnesium sulfate as it can damage the chemical bonds and intermolecular bonding of constituents and hydrated mass of the cement paste.

Concrete cubes were prepared containing varying PCMPW fractions and types. Cubes were cured for 28 days with normal water curing. After the curing period, cubes were cleaned, dried, and the weights were recorded. The specimens were placed in a sulfate solution prepared with 10% sodium sulfate. After 60 days of sulfate curing, the cubes were cleaned and weighted to obtain the weight difference.

The objective of the test was to assess the effects of addition of PCMPW flakes in concrete and to check the possible changes in the resistance against the sulfate attack. It was observed during the test that the presence of the PCMPW flakes did not have any significant negative effects on the properties (Figure 5.17).

Resistance to acid solution attack on concrete was determined by immersing the specimens in acid solution. Acid in solution form attacks the hydrated cement paste compounds and chemically converts them into new compounds that are soluble in water. Consequently, when concrete is in contact with acid solution, leaching in concrete takes place and concrete strength reduces (Figure 5.18).

FIGURE 5.17   Effect of chemical on specimen surface.

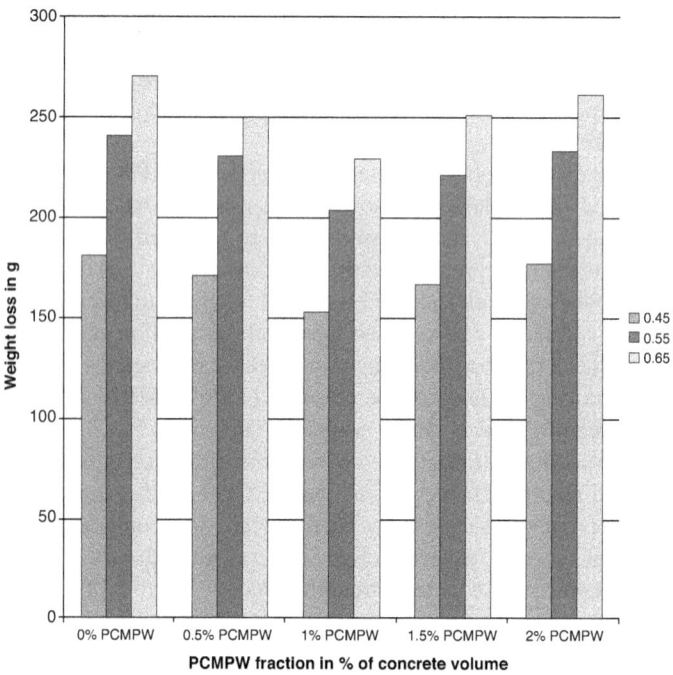

FIGURE 5.18   Effect of PCMPW fiber variation on acid resistance of concrete.

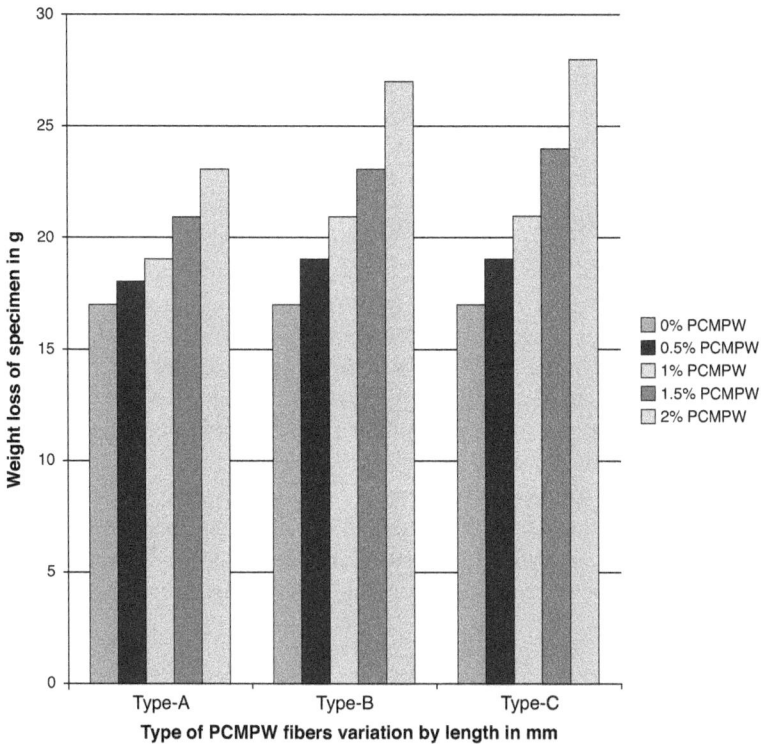

FIGURE 5.19 Effect of PCMPW type on acid resistance of concrete.

5.5.4 EFFECT OF **PCMPW** FRACTION

Acid resistance of concrete containing PCMPW was measured by obtaining weight difference of specimens before and after acid curing. The results are shown in Figure 5.19. The weight loss reduced in the range of 5%–10% at 0.5% and 8%–20% at 1% addition of PCMPW for a given PCMPW type and water cement ratio. For the increased PCMPW dosage at 1.5% and 2%, the weight difference was larger than the 1% dosage of PCMPW. The loss of weight of the specimen before and after the test was increased by 4%, 10%, and up to 15% for given water cement ratio with varying PCMPW type for the concrete containing 1.5% and 2% of PCMPW.

5.5.5 EFFECT OF **PCMPW** TYPE

Figure 5.19 describes the effect of PCMPW type on acid resistance of concrete. Type A and type C demonstrated nearly similar response to acid resistance, while type B PCMPW showed relatively distinct behavior for a given water ratio and fraction of PCMPW. The measured reduction in weight difference was in the range of 5%–15% at lower dosage of PCMPW for given water cement ratio in case of type A and type C PCMPW. The rate of weight reduction decreased, and more weight loss was observed beyond 1% of PCMPW dosage for all three types of plastics. However, the loss of weight was significant in concrete containing type B PCMPW in all three mixes beyond 1% dosage. The weight loss increased up to 14% at 2% with the addition of type B PCMPW in concrete. Type A and type C PCMPW resisted the effects of acid attack on concrete effectively than type B at low dosage and for given water cement ratio.

5.5.6 EFFECT OF WATER CEMENT RATIO

Figure 5.20 shows that water cement ratio of 0.45 showed better resistance to acid attack by reducing the weight loss due to the acid ingress in the specimen. Weight loss was significant for concrete prepared with 0.65 water cement ratio. Specifically the mix containing type B of PCMPW showed larger weight reduction at higher volume contents beyond 1% as shown in Figure 5.20. Reduction in weight loss was found to be 14% in case of mixes prepared with w/c of 0.45, while the rate of weight reduction reduced with increased w/c and showed weight loss up to 20% for concrete prepared with higher values of water cement ratio, namely 0.55 and 0.65.

PCMPW flakes acted as blocking contents to acid ingress and reduced the corrosive chemical process with cement paste. The benefit of random distribution of flakes provided excellent results against the surface damage too. PCMPW in varying

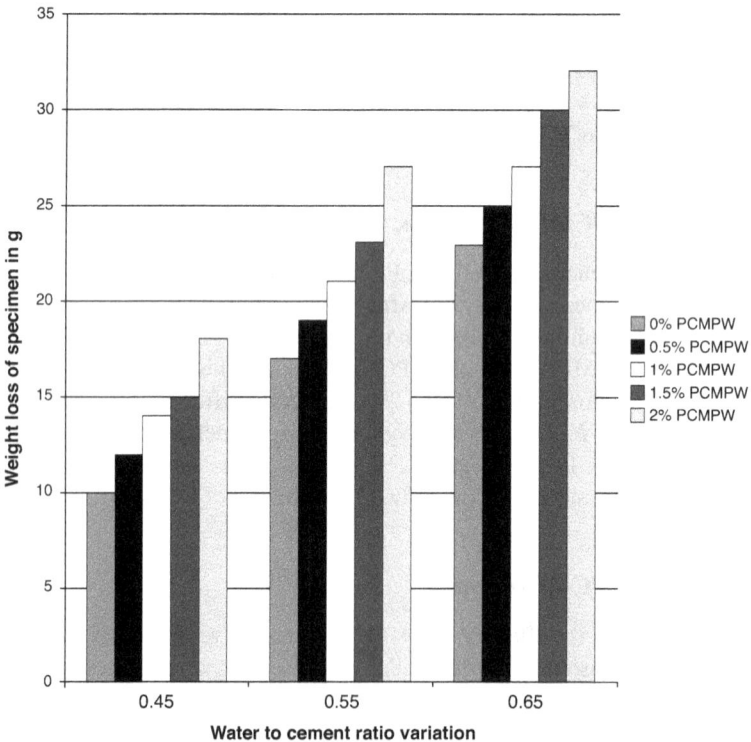

FIGURE 5.20   Effect of water cement ratio on acid resistance.

types and fractions effectively reduced the acid ingress through the void structure of hardened concrete. As the acid promotes the leaching of paste and loss of strength, the presence of PCMPW reduced the leaching of material from the mass.

The presence of sulfate contents in water or soil creates spalling of concrete by reacting with the calcium and lime contents available in hardened concrete. The leaching of material from the hardened mass of concrete has been generally found where the sulfates interact through surface erosion, resulting in further ingress inside the mass. The effect of addition of PCMPW in concrete for sulfate resistance was measured by measuring the weight loss of the specimen during the curing period.

5.5.7   EFFECT OF PCMPW FRACTION

The rate of weight loss increased with the increased PCMPW fractions. From 0% to 2% addition of PCMPW in concrete, the weight loss increased from 4% to 29%. However, concrete prepared with 0.45 water cement ratio showed resistance to the sulfate attack by reducing the weight loss up to 10% at lower dosage of 1% of PCMPW in mixes prepared with 0.45 water cement ratio. Beyond 1% dosage of PCMPW the weight loss was significant for the given water cement ratio and PCMPW type.

5.5.8   EFFECT OF PCMPW TYPE

The test results showed that the change in PCMPW type did not affect the sulfate resistance of concrete. The response of concrete containing varying types of PCMPW for given water cement ratio and fraction showed nearly similar response. Type A of PCMPW showed effective resistance to the sulfate attack on concrete compared to other types. The weight loss of concrete was less than 10% for all mixes containing type A of PCMPW. Type B and type C exhibited more weight loss and at a uniform rate of increase. Out of all three types of PCMPW, type B and type C showed larger weight difference up to 29% for the dosage of 2% of PCMPW. Inclusion of PCMPW in concrete did not significantly resist the sulfate attack. However, PCMPW type A reduced the weight loss that occurred during the sulfate curing of concrete.

5. 5.9   EFFECT OF WATER CEMENT RATIO

The weight loss increased with an increase in water cement ratio values from 0.45 to 0.65. The water cement ratio of 0.45 effectively resisted the sulfate attack for the given PCMPW type and fractions.

5.5.10   OBSERVATIONS FROM THE ACID AND SULFATE ATTACK RESPONSE OF CONCRETE

Results of acid and sulfate attack test demonstrated the relationship of the addition of PCMPW flakes and surface degradation of concrete due to the direct contact of the solution. It was observed that the mixes prepared with lower water cement ratio and possessing higher strength were found to be suitable for further experimentation on the chemical attacks on concrete containing the PCMPW flakes.

It was noticed that the presence of PCMPW flakes did not respond to the surface degradation due to the acid and sulfate solution. This could be explained as the presence of PCMPW flakes reduced the micro level cracks in the hardened mass and resisted the solution ingress. On the other hand, the flakes were mixed and highly scattered inside the mass rather than remaining present on the surface. This is an important observation from the surface reaction of concrete toward the environmental degradation perspective.

It was noticed that water cement ratio as a test variable remained unaffected by the varying size and contents of the flakes. Therefore, further durability tests pertaining to the chemical attacks like chloride penetration were conducted at a constant water cement ratio of 0.45 and the investigation was focused on the effects of other variables, namely size and contents of the PCMPW flakes. The readers can refer to the details provided in the published work by the author as per the references [1,2].

5.6   STRUCTURAL PERFORMANCE OF MEMBERS PREPARED WITH A MODIFIED CONCRETE

Structural response of concrete containing PCMPW was evaluated by obtaining stress–strain relationship of concrete in axial compression and by the flexural response of reinforced concrete beam specimens containing varying PCMPW fractions.

The test results of workability, strength, and durability properties were used to obtain the material response of concrete containing PCMPW. The effects of PCMPW on material behavior were analyzed by studying the alterations in the properties for varying PCMPW fraction and sizes along with the varying water to cement ratio of mixes.

It is equally important to study the structural behavior of a member containing PCMPW as the application of the new material to obtain the changes in the standard response of the members. To what extent the standard structural response could be affected by PCMPW was assessed by performing flexure tests.

Careful observation of test results of workability, strength, and durability property revealed that among all three types of PCMPW, namely type A, type B and type C, the addition of type A showed minimal alterations in the properties with respect to the reference concrete. The alterations in the material properties were found in the range of 5%–15% only.

However, such results were obtained up to the fraction of 1% by volume of PCMPW. Beyond 1% dosage of PCMPW, the properties were reduced. These results were common to all three mixes though; the structural concrete should be prepared with water to cement ratio not more than 0.45 as a general practice. Therefore, the structural response was studied by PCMPW type A in concrete prepared with 0.45 as water to cement ratio.

5.6.1   Stress–Strain Relationship

Mixes and casting of specimens

The concrete mix was prepared with 0.45 as water to cement ratio and the mix design was carried out to obtain the quantities of constituents. PCMPW of type A was added in the fraction range from 0% to 2% by volume of the concrete mix.

FIGURE 5.21   Stress–strain test specimens.

The stress–strain relationship of concrete was assessed on cylinder specimens as per the guidelines of IS: 516-1999. For every cylinder three cubes were required to obtain the average strength. Therefore, cubes of standard dimensions were prepared with the same concrete used in casting of cylinders of 150 mm diameter and 300 mm height. The specimens are shown in Figure 5.21.

5.6.2   TESTING OF SPECIMENS

Molds were properly conditioned and prepared as per the guidelines of IS: 516-1959. Cube strength as the average of three specimens was considered for load and strength calculation as per the guidelines of the code. The cylinders and cubes were cured for 28 days at room temperature. Cylinders were subjected to axial compressive load by CTM of 2,000 kN load capacity.

The test method explained in the code primarily dealing with the investigation of modulus of elasticity of concrete by extensometer was used to determine the strain values corresponding to the applied stress. The rate of loading was maintained at 140 kg/cm^2/min. During the tests, the failure patterns, recording of strain at different stages, values of final stress at failure strain, and peak stress values were obtained and recorded (Figure 5.22).

FIGURE 5.22   Tested concrete specimen for stress–strain relationship.

5.6.3   OBSERVATORY NOTES

Peak stress and deformation response at corresponding strain required uniform slow loading rate. Effects of addition of PCMPW on stress–strain relationship were assessed by observing the crack patterns and failure of the specimens. While performing the tests it was observed that the reference concrete containing 0% PCMPW showed abrupt loss of strength and showed brittle failure beyond the peak stress values. Concrete containing PCMPW flakes showed controlled cracking at peak stress values and beyond the peak stress, the specimens exhibited comparatively less brittle response. Stress and corresponding strain values were recorded for each specimen and converted to graphical format to obtain the stress–strain curve.

An analytical equation from each stress–strain curve including the reference concrete was retrieved. A standard analytical equation was used to validate the experimental results. The stress–strain curve obtained from the tests on reference concrete is shown in Figure 5.23. The compressive strength of concrete reduced with the increased PCMPW fraction. This response of concrete to axial deformation was reflected by the curve profiles at the peak stress. The curve profiles were sharp at peak stress for the reference concrete and became flat in case of concrete containing increased dosage of PCMPW.

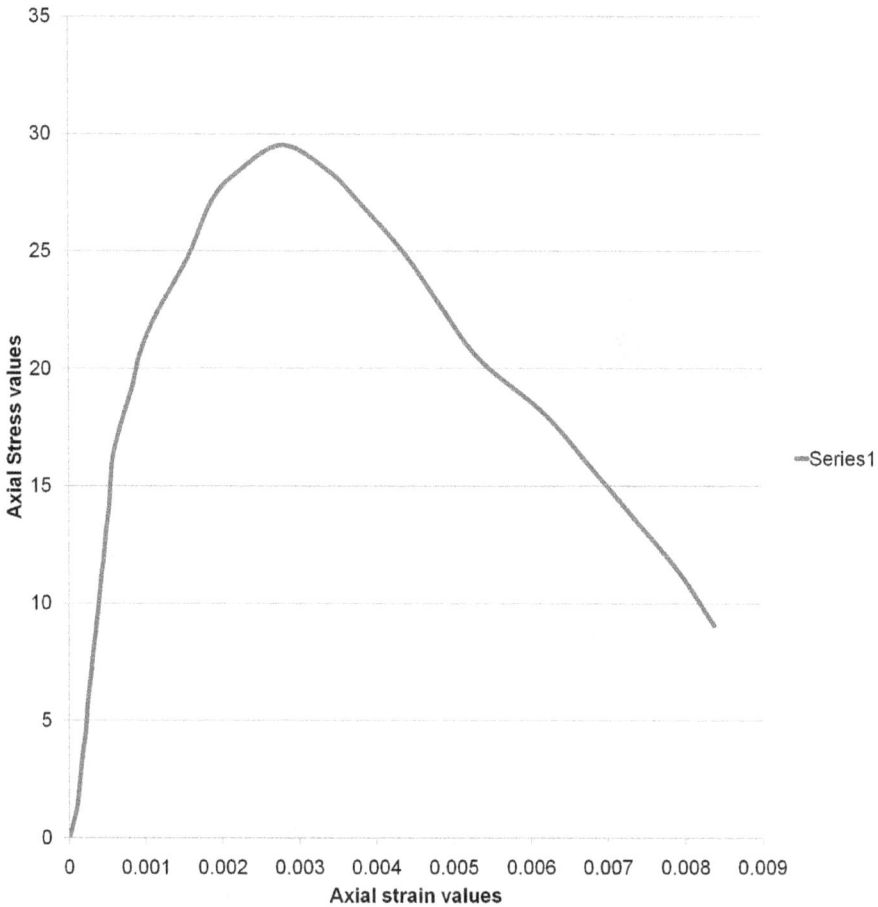

FIGURE 5.23 Stress–strain curve of modified concrete.

5.6.4 BEAM SPECIMENS

Flexural response of concrete containing PCMPW was obtained by performing a four-point flexural load test on reinforced beam specimens. A total of 12 beams were prepared containing varying PCMPW flake fractions of type A flakes including the reference concrete. The beams were prepared with steel reinforcement with varying steel proportions and designed as balanced sections, under-reinforced section, and over-reinforced section. The beam mold dimensions were 150 mm × 200 mm × 1,500 mm. The reinforcing steel cages were first prepared and placed in the beam molds with sufficient cover and all sides of the molds were oiled properly before casting. The beams were cured for 28 days in the laboratory at room temperature. To obtain clear failure patterns on the beam surfaces, all sides of beams were painted with white color (Figure 5.24).

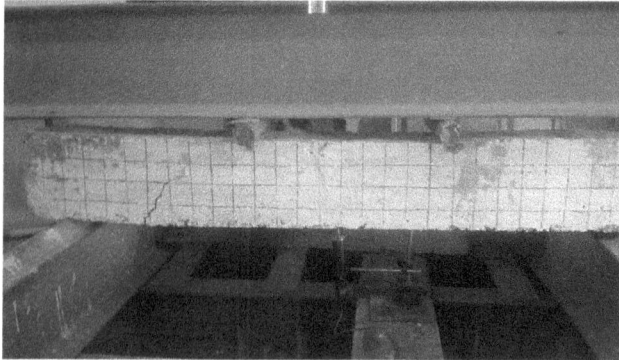

FIGURE 5.24 Beam specimens for flexure test.

5.6.5 EXPERIMENTAL SETUP AND TEST

The beam specimen was mounted over a load frame of 500 kN load capacity. A complete experimental setup is shown in Figure 5.25. The objective of the tests was to obtain relationship between load and corresponding deflection and to study the flexure response of beams up to the failure. The central deflection of beams was measured by a dial gauge placed at the bottom. Load values were recorded manually for the first crack load and final crack load. Ratio of deflection caused by first crack load to final crack load was regarded as the ductility index for a beam. The ductility index varied with the proportion of steel reinforcement and PCMPW fractions.

Experimental moment carrying capacity of a beam section was obtained by the load and perpendicular distance from the end support of the beam. In addition, moment-resisting capacity was theoretically calculated by the modified stress–strain

FIGURE 5.25 Tested beam under flexure actions.

equation and the Hognestad equation. A comparative study was carried out on all the three values of moment-resisting capacity for all beams. The differences of experimental and analytical values were used to assess the effects of addition of PCMPW.

Crack patterns developed on beam surfaces were carefully observed. Crack patterns were used to describe the nature of failure of beam section.

The structural response of concrete was evaluated by preparing cylinders and reinforced beams. Cylinders were subjected to axial compressive loads and stress–strain relationship was obtained. Reinforced concrete beams with three varying reinforcing steel proportions were subjected to four-point loading conditions in the load frame. Stress–strain relationship was studied by retrieving the stress–strain curves and comparing them with standard curves of reference concrete. Analytical model suggested by Hognestad was utilized for comparison. A modified analytical model was also proposed based on the experimental stress–strain curves.

Reinforced beams were studied for altered moment-resisting capacity for varying PCMPW flake fractions of type A for balanced, under-reinforced, and over-reinforced sections. Theoretical and experimental values were compared for the assessment of standard response of beams. Moreover, load to deflection relationship was obtained for all beams including the reference beams containing 0% PCMPW flakes. The objective was to evaluate the alterations in the standard response of concrete when used in structural member applications.

5.6.6 Stress–Strain Relationship

5.6.6.1 Stress–Strain Curves: Mixture 1

Concrete mixture 1 was prepared with water to cement ratio 0.45. Type A flakes were mixed in the concrete ranging between 0% and 2% with 0.5% increment in dosage like the material evaluation.

Each cylinder was cured for 28 days before testing and mounted on CTM used for compression tests and employed with an extensometer. The IS: 516-1959 was referred for conducting the tests to obtain stress–strain values. Stress–strain curves are shown in Figure 5.26.

5.6.6.2 Observations

Stress–strain curves provide information about the deformation characteristics of concrete containing varying PCMPW flake fractions. As discussed earlier, type A flakes were used in all tests.

The addition of PCMPW reduced the strength of concrete mix; however, it also increased the crack resistance of the concrete with increased fraction up to 1%.

Failure of specimens was changed due to the addition of PCMPW specifically beyond the peak stress values.

The strain values corresponding to the peak stresses increased with an increase in the PCMPW flakes. However, the trend was observed till 1% of dosage. Beyond 1% of dosage, significant reduction in the stress was observed.

It is seen that the curve exhibits smooth lines containing less strength. This trend continues with increased plastic contents.

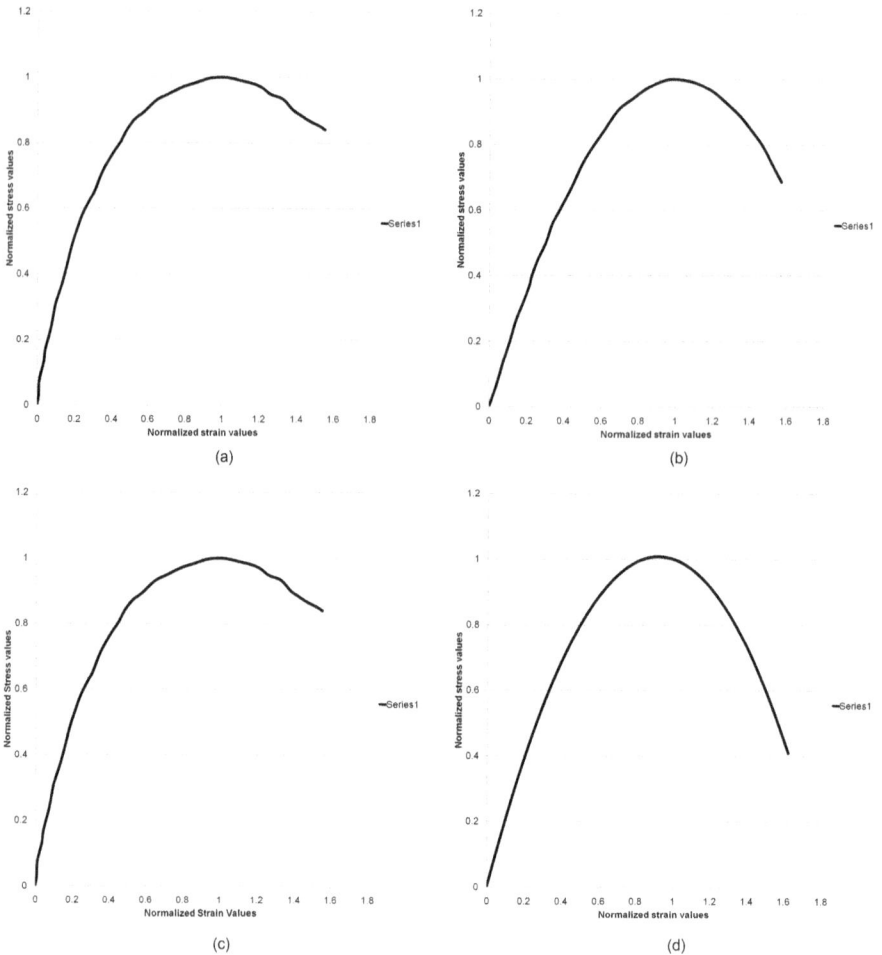

FIGURE 5.26 (a) Stress–strain curve for reference concrete with normalized values. (b) Stress–strain curve for concrete with 0.5% PCMPW flakes. (c) Stress–strain curve for concrete with 1.0% PCMPW flakes. (d) Stress–strain curve for concrete with 1.5% PCMPW flakes of concrete mixture 1 containing varying PCMPW fractions.

It was observed while performing the experiments that the reference concrete failed catastrophically, after peak stress values.

The failure response of modified concrete with PCMPW showed more control and failed less catastrophically.

5.6.6.3 Stress–Strain Curves: Mixture 2

Mixture 2 was prepared with 0.55 as water to cement ratio. Cylinders were prepared containing varying fractions of PCMPW flakes of type A. All the test methods were similar as in case of mixture 1. The results were used to generate stress–strain curves.

5.6.6.4 Observations

Concrete prepared with 0.55 water cement ratio exhibited reduced strength and reduced peak load values before failure of the specimens. It is observed that as a general response of concrete to axial compression at higher water to cement ratio, concrete failure is more brittle. The patterns of cracks on the surfaces are wider and more inclined toward the upper portion of cylinders. Inclusion of PCMPW further extended this collapse mechanism and increased the fraction of PCMPW, resulting in more brittle nature of the mixture.

5.6.6.5 Stress–Strain Curves: Mixture 3

Mixture 3 was prepared with 0.65 as water to cement ratio. The mixes prepared with higher values of water to cement ratio are generally preferred for ready mix concrete, Shotcrete concrete, and lean mixes to be used for the leveling of nonstructural parts of structure.

5.6.6.6 Observations

Concrete subjected to axial compression prepared with high water to cement ratio showed continuously reducing strength of concrete. Increased water content promotes free or excess water in the mix. This increases the voids in the concrete and reduces strength. Addition of PCMPW up to an extent occupied the air and water voids; however, they do not participate in the strength gaining mechanism and therefore do not improve the strength.

Structural members, namely beams, are affected by combined effects of forces in actual practice. The members generally respond to the actions according to the strength of material to respective load actions. Beams are primarily subjected to bending and shear deformation. The structural response of beams to such actions is governed by the strength of concrete in flexure and shear.

Concrete modified with PCMPW was tested for the structural response of material in flexure. The beam sections were prepared with reinforcing steel and concrete containing varying fractions of PCMPW. Important observations namely first crack load and final crack loads were recorded. Deflection values corresponding to load values were measured with dial gauges. Crack patterns were observed carefully in each beam after and during the tests. Experimental values were compared with the analytical values obtained by employing standard model and proposed model from the stress–strain curves.

5.6.7 Response by Balanced Section

Beams with balanced steel reinforcing sections were prepared containing 0%–1.5% volume fractions of PCMPW of type A. Beams with 0% flakes failed in shear as shown in Figure 5.27.

The response of beams containing 0.5%, 1%, and 1.5% of flakes by volume of the mix was governed by flexural cracks. The crack patterns were found to be linear for the beams containing increased fractions of PCMPW.

It was observed that the first crack developed at lower value of load compared to the reference beam containing 0% flakes. The scenario was similar for all beams with increased flake contents. Ratio of deflections at final crack load to first crack

FIGURE 5.27   Shear failure of balanced beam section with 0% of flakes.

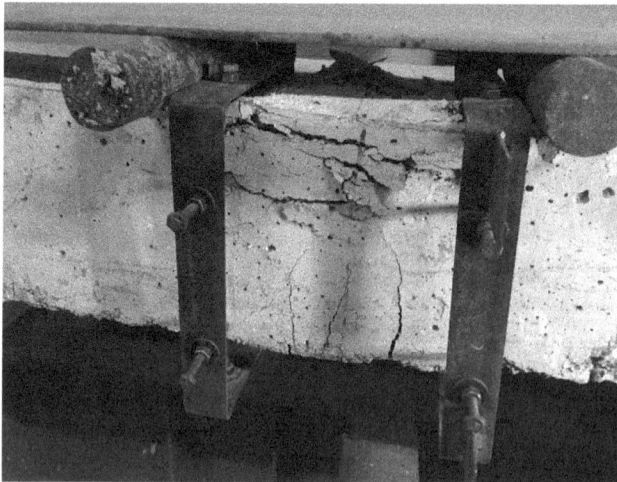

FIGURE 5.28   Flexural failure of balanced section with 1% of PCMPW flakes.

load was obtained for all beams and regarded as the ductility index for the beams. The ductility index was the maximum for 1% volume of PCMPW addition. It was observed that the increased flake contents reduced the ductility index of the beam. A relatively mixed response dominated by flexural failure of the beam was observed in the beam as shown in Figure 5.28 at 1% of volume fraction.

However, the failure pattern was governed by the percentage of PCMPW up to the initial loading only. Fine cracks were developed at the lower load values compared to the reference beams containing 0% PCMPW. This response was confirmed by the experimental values obtained from the cubes subjected to compressive loading. Cubes containing higher dosage of PCMPW showed reduced strength. A very similar response was observed in the beams containing varying fractions of PCMPW.

5.6.8 RESPONSE BY UNDER-REINFORCED SECTION

Under-reinforced beam sections were prepared by addition of varying PCMPW fractions. The initial and final cracks were developed at the lower load values compared to the balanced sections for all fractions of PCMPW. Similarly, the deflection values were also larger than that of the balanced sections. With the increased PCMPW fraction values, the load values further reduced, and the deflection of beam section increased. All beam specimens failed in flexure accompanied by secondary shear cracks near the supports (Figure 5.29).

The structural response of beam sections, namely balanced and under-reinforced sections, differed for the deflection values and loads at initial and final failure. The under-reinforced sections exhibited larger deflections at lower load values and showed increased values of the ductility index compared to the balanced sections at a constant PCMPW dosage. Beyond 1% dosage of PCMPW flakes, the deflection continued to increase with reduced margin. At 1.5% dosage of PCMPW, the section showed negligible load carrying capacity and failure of compression and tension zone was significant.

5.6.9 RESPONSE BY OVER-REINFORCED SECTION

Over-reinforced sections exhibited shear failure for all fractions of PCMPW. The flexural response was governed by shear cracks near supports propagating to the point of application of loads. As shown in a figure the beam showed shear failure irrespective of the PCMPW contents.

The sections resisted maximum loads and showed minimum deflection at mid span. The ductility index was also reduced compared to the balanced and under-reinforced sections. The primary reason for this response is the availability of steel reinforcement at the section responsible for resisting the deformation in flexure. It was observed that the standard response by over-reinforced sections was not affected due to the presence of the PCMPW in concrete.

The over-reinforced sections containing PCMPW showed standard structural response in flexure as each beam specimen failed in shear. However, the crack

FIGURE 5.29 Flexural failure of under-reinforced section with 0.5% of PCMPW flakes.

patterns showed little influence of inclusion of PCMPW. The unconfined concrete showed linear and controlled cracks at initial loading and gradually propagated from tension zone toward compression zone.

5.7   SPECIAL REMARKS ON THE USE OF PLASTIC WASTES IN CONCRETE

Summary of results indicated several important attributes on the use of PCMPW in concrete from the material response perspective. The results were carefully analyzed with three variables, namely water cement ratio, PCMPW fraction, and PCMPW type.

5.7.1   Effect of Water Cement Ratio

Varying water cement ratio from low to high values, namely from 0.45 to 0.65, showed reduction in compressive strength of concrete mixes for given PCMPW type and fraction.

Mixes prepared with higher water contents showed less reduction in fresh properties at a given volume fraction of flakes and sizes. It can be observed that more water content in the mix provided improved workability even with the presence of the PCMPW.

The overall response of all mixes to particular property was as per the standard response of the material like reference concrete. This confirmed that water cement ratio can be adapted according to the need of strength expected from concrete, for any fraction of PCMPW and size.

5.7.2   Effects of PCMPW Fraction

Increased PCMPW fraction showed reduction of fresh properties and compressive strength. The lower volume fraction showed relatively insignificant reduction, however. Similar trend was observed in durability response of concrete.

At constant water cement ratio and size of flakes, up to 1% dosage of PCMPW is found suitable to be mixed in concrete. Reduction in a particular property remained within the upper limit of 15%.

5.7.3   Effect of PCMPW Type

Type A PCMPW was the most preferable size of flakes to be used in concrete for all test conditions. Small size did not affect strength gaining mechanism or loss of strength significantly. The standard structural response of concrete was studied with PCMPW type A.

Type B was found to be a non-preferable size as it showed maximum reduction among all three types.

Type C flakes showed excellent crack resistance and impact resistance and were found suitable for all test conditions up to the optimum dosage.

Plastics irrespective of types and forms are difficult to dispose of. Across the world, landfill has been considered as the main option due to associated complications of municipal waste containing plastics. Recycling and reuse of plastic waste have not attained an acceptable rate of disposal due to the requirement of large human input and high energy consumption. Therefore, innovative ways of disposal of plastic waste have become highly desirable.

The present study dealt with the concept of producing green concrete by utilizing post-consumer metalized waste plastics as one of the constituents for availing the dual benefit of reducing the energy needs for concrete manufacturing and alternative for waste plastic disposal. Plastics manufactured with varying resins exhibit different characteristics, properties, and responses when used in concrete. Use of metalized waste plastic was required to be thoroughly examined for its impacts on concrete response and behavior for such reasons.

Extensive experimental investigations were employed to obtain workability, strength, and durability response of concrete modified with PCMPW. The standard structural response of beam members in flexure was also investigated for the possible changes due to addition of PCMPW in concrete. The following are the conclusions derived based on the experimental results [2-4].

Experimental investigations showed the possibility of preparing green concrete by utilizing PCMPW, providing an alternative to the landfill disposal methods for protecting the environment from plastic waste pollution.

The addition of PCMPW of 1 mm size flakes at a dosage of 1% by volume of the concrete mix showed negligible effects on fresh and hardened properties of concrete. This quantity shows good agreement to the objective of mitigating littering/waste management problem of plastic bags and producing green concrete.

Some mechanical response of concrete containing PCMPW was also found even better than ordinary concrete. It showed better crack resistance and higher energy absorption under impact loading.

Inclusion of waste plastic flakes improved the durability of concrete compared to plain concrete. Concrete containing PCMPW showed reduction in permeability for water and gas ingress. It also showed better resistance to acid, sulfate, and chloride attacks at the optimum dosage of PCMPW [1].

Standard structural response of reinforced concrete beams containing PCMPW was satisfactory. Ductility of beams and strain capacity of unconfined concrete in post-peak stress region under axial compression were found to be improved owing to the addition of PCMPW. The readers are encouraged to read referenced [2-4] for further reading on the topics related to the utilization of PCMPW in concrete composites.

REFERENCES

[1] Bhogayata, Ankur C., and Narendra K. Arora. "Fresh and strength properties of concrete reinforced with metalized plastic waste fibers." Construction and Building Materials 146 (2017): 455-463.

[2] Bhogayata, Ankur, and N. K. Arora. "Feasibility study on usage of metalized plastic waste in concrete." Contemporary Issues in Geoenvironmental Engineering: Proceedings of the 1st GeoMEast International Congress and Exhibition, Egypt 2017 on Sustainable Civil Infrastructures 1. Springer International Publishing, 2018.

[3] Bhogayata, Ankur C., and Narendra K. Arora. "Impact strength, permeability and chemical resistance of concrete reinforced with metalized plastic waste fibers." Construction and Building Materials 161 (2018): 254-266.

[4] Bhogayata, Ankur C. "Concrete reinforced with metalized plastic waste fibers." Use of Recycled Plastics in Eco-efficient Concrete. Woodhead Publishing, 2019. 349-367.

6 Material Properties of Concrete Prepared with Industrial Wastes

6.1 MIX DESIGN OF THE CONCRETE

To develop the geopolymer concrete technology, a rigorous trial-and-error process was used. The focus of the study was to identify the salient parameters that influence the mixture proportions and the properties of low-calcium fly ash-based geopolymer concrete. In the current manufacture of ordinary Portland cement, concrete is used. The aim of this action was to ease the promotion of geopolymer concrete as a "new" material to the concrete construction industry. Geopolymer concrete can be made using various source materials; the present study used only ASTM class F fly ash. Also, as in the case of ordinary Portland cement, the aggregates occupied 75%–80% of the total mass of concrete. To minimize the effect of the properties of the aggregates on the properties of fly ash-based geopolymer, the study used aggregates from a single source. The materials utilized for making geopolymer concrete were low-calcium (ASTM class F) fly ash, alkaline liquid (sodium silicate and sodium hydroxide), aggregates, water, and super plasticizer. In the present investigation, fly ash containing low calcium (<10% CaO as per American Society for Testing and Materials) C618 class F was used as the source material. The chemical compositions of the fly ash were determined by X-ray fluorescence (XRF) analysis as given in Table 6.1. It can be seen from Table 6.1 that the fly ash contained a very low percentage of carbon. In this composition, the molar Si-O-Al ratio was about 2, and the calcium oxide content was very low. The iron oxide (Fe_2O_3) contents are

TABLE 6.1
Chemical Composition of Class F Fly Ash

Oxides	(%) by Mass
SiO_2 as silica	54.70
Al_2O_3 as alumina	23.55
Fe_2O_3 as ferric oxide	8.89
CaO as calcium oxide	2.19
MgO as magnesium oxide	1.31
K_2O as potassium oxide	0.75
SO_3 sulfur trioxide	0.66
Na_2O as sodium oxide	0.35
Loss of ignition	1.63

DOI: 10.1201/9781032621340-6

relatively high. Specific gravity of fly ash was 2.28. The other materials namely fine and coarse aggregates were utilized in the program as available from the local resources; fine river sand with fineness modulus of 2.78 is used as a fine aggregate. The necessary tests were carried out on the samples as per IS: 2386-1968(III) code guidelines. The crushed coarse aggregates of granite type were utilized by the mixtures in the experimental works. A good-quality, well-graded coarse aggregate of size 10 and 20 mm was used in the preparation of all test specimen.

6.1.1 ALKALINE SOLUTION

The alkaline solution was a combination of sodium silicate and sodium hydroxide. Sodium-based solutions were chosen because they were cheaper than potassium-based solutions. The sodium hydroxide solids were of a technical grade in flake form (3 mm), with a specific gravity of 2.21, 98% purity. The sodium hydroxide (NaOH) solution was prepared by dissolving either the flakes or the pellets in water. In this experimental work sodium hydroxide was used in three different concentrations such as 8 M, 12 M, and 16 M. The chemical composition of the sodium silicate solution was $Na_2O = 15.87\%$, $SiO_2 = 31.73\%$, and water 52.4% by mass. The other characteristics of the sodium silicate solution were specific gravity $= 1.62$ g/cc and viscosity at $20°C = 400$ cP. To improve the workability of the fresh geopolymer concrete, high-range water reducing naphthalene sulfonate super plasticizer was utilized for all mixes.

It is suggested that before conducting the full-length experimental work, pilot studies are necessary. The objective of preliminary investigation is to find the proportions in which the geopolymer concrete materials like fly ash, alkaline liquid, water, fine aggregate, and coarse aggregate should be combined to provide the specific strength workability and durability. The ingredients of concrete are obtained by the required performance of concrete in the plastic and the hardened stages. Concrete must be of satisfactory quality in both stages, namely fresh stage and hardened stage. This can be best practiced by trial mixes. During trial mixes, different sodium silicate solution as alkaline activator was used in geopolymer concrete to check the feasibility of material used in concrete mix and the curing temperature, curing time, and rest period after hot oven curing. For instance, for the work being discussed, in the preliminary study we cast nine cubes per mix. Then workability test on fresh concrete and compression test on hardened concrete were performed. In our study we were only varying parameters such as sodium silicate to sodium hydroxide ratio, molar content of sodium hydroxide, alkaline liquid to fly ash ratio, curing temperature, curing time etc. With using constant parameters such as fly ash to alkaline liquid ratio and adding extra water, the compressive strength of geopolymer concrete for all mixture proportions was measured. Based on the trial mixes, the following mix designs were prepared.

For each mixture proportion, the concrete cube of various proportions of sodium hydroxide molarity, curing temperature, and curing time was prepared. In Table 6.2, sodium silicate to sodium hydroxide ratio was 2 and 3. We cast cubes by taking these two ratios with different parameters like sodium hydroxide molar content (8, 12, and 16 M), curing temperature (60°C, 80°C, and 100°C), curing time (24 and 48 hours), and rest period (3, 7, and 28 days) after hot oven curing.

TABLE 6.2

Mix Design for GPC Based on the Trial Mixes

Constituents	Unit	Mixture I	Mixture II
Fly ash	Kg/m³	368	368
Fine aggregate	Kg/m³	554.4	554.4
Coarse aggregate			
10 mm	Kg/m³	443.52	443.52
20 mm	Kg/m³	850.08	850.08
NaOH solution	Kg/m³	46	61.33
Na₂SiO₃ solution	Kg/m³	138	122.66
Extra water	Kg/m³	29.44	29.44

6.1.2 MIXING

In geopolymer concrete, the sodium hydroxide solution and the sodium silicate solution were mixed at least 1 day prior to use as alkaline liquid. In this solution, the sodium hydroxide flakes were dissolved in distilled water to make NaOH solution. The sodium hydroxide solution was kept at 8, 12, and 16 M for each trial mix. First, coarse aggregate, fine aggregate, and fly ash were mixed in an electric tilting drum mixer machine for about 3–4 minutes. After this dry mixing, the alkaline solution, super plasticizer, and water were added in the drum and mixed for 4–5 minutes for proper wet mixing. In this preliminary study, it was found that the specimens were disturbed during decoupling. Geopolymer concrete was stuck in the inner face of specimens.

6.1.3 CURING TRIALS

After casting of geopolymer concrete specimens, it should be cured at elevated temperatures in the oven. To prevent excessive evaporation, specimens should be wrapped with a plastic bag and a lid placed on a small specimen mold; a suitable method was needed for large size geopolymer concrete specimens. In the trial mix, the curing temperature was 60°C, 80°C, and 100°C, kept at the oven for 24 and 48 hours. A total of nine cubes were cast per batch. In nine cubes, three cubes were kept at 60°C, three cubes were at 80°C, and other three cubes were at 100°C in a hot air oven. After curing the specimen at elevated temperature, it should be kept for different rest periods for compressive strength. The compressive strength was measured after 3-, 7-, and 28-day rest period.

6.1.4 THE TEST RESULTS FOR PRIMARY DESIGN

Specimens for strength tests in compression should be made from all trial batches and from several batches after a satisfactory mixture has been established to determine if the strengths are within the range intended.

TABLE 6.3

Mix Design for GPC Consisting of Varying Proportions of Fly Ash

Constituents	Unit	Mixture I	Mixture II
Fly ash	Kg/m³	380.69	356.13
Fine aggregate	Kg/m³	554.4	554.4
Coarse aggregate			
10 mm	Kg/m³	443.52	443.52
20 mm	Kg/m³	850.08	850.08
Fly ash/alkaline liquid		0.45	0.55
Na₂SiO₃ solution/NaOH solution		2 and 3	2 and 3
Silicate oxide/sodium hydroxide		2.1	2.1
NaOH concentration	M	8, 12, and 16	8, 12, and 16
Extra water	Kg/m³	30.45	28.49

The results showed that by increasing the temperature from 60°C to 100°C, the compressive strength increased by 25%–50% for constant alkaline/fly ash ratio, Na_2SiO_3/NaOH ratio, and NaOH molarity. And, it was observed that there was no effect of rest period on compression strength of geopolymer concrete. By increasing the NaOH concentration from 8 to 16 M, the compressive strength also increased. Based on the studies, it may be suggested that the mix design should be made by keeping the molar content in focus. Table 6.3 gives the mix design proposed for preparing the final mixtures of GPC containing industrial wastes.

6.2   METHODS OF MANUFACTURING

The GPC concrete may be prepared in a variety of ways. However, to provide a simple explanation on the making of the mixtures, a direct approach has been presented in this section. The method of making the GPC is largely similar to that of cement-based concrete. The GPC also consists of two portions: dry contents and a solution instead of water. The dry component may be prepared with fly ash, sand, and aggregates. The wet component includes the making of the alkaline solution using different proportions of sodium silicate and sodium hydroxide. The sodium silicate as a liquid also consists of varying proportions and ratios of silicate to sodium oxides. This ratio may be specified to the manufacturers and the desired mixture of sodium silicate may be obtained. The typical steps of manufacturing the GPC mixes include: (a) alkaline solution preparation, (b) mold preparation, (c) batching of the materials, (d) mixing, (e) casting, and (f) vibrating. It is to be noted that the steps are similar to conventional concrete making; however, the alkaline solution should be prepared 1 day prior to the day of casting and mixing.

The sodium hydroxide (NaOH) flakes were dissolved in distilled water to make sodium hydroxide solution. The NaOH solution was kept at 8 and 12 M for trial mix. The sodium silicate solution and the sodium hydroxide solution were mixed 1 day prior to use to prepare the alkaline liquid. On the day of casting of the specimens, the super plasticizer was mixed with the extra water to prepare the liquid component of the mixture. Alkaline solution along with super plasticizer and extra water was added to dry mix and mixing was done for about 2–3 minutes to achieve workable concrete mix. It was found that the amount of water in the mixture played an important role on the behavior of fresh concrete. When the mixing time was long, mixtures with high water content bled and segregation of aggregates and the paste occurred. This phenomenon was usually followed by low compressive strength of hardened concrete.

In this pilot study, we observed that the specimens had problems during demolding. Concrete was stuck to the inner face of the specimens. Therefore, we stuck white cello tape on the inner face for all the molds and the specimens were found in good condition even after hot air oven curing. From the pilot study, it was decided to adopt the following standard process of mixing in all further studies. Mix sodium hydroxide solution and sodium silicate solution together 1 day prior to casting of geopolymer concrete. Mix all dry materials in the electrical tilting drum mixer machine for about 3 minutes. Add the polymer plastic and again dry mix it for a further 2 minutes. Add the liquid component of the mixture at the end of dry mixing and continue the wet mixing for another 2–3 minutes to get workable geopolymer concrete.

6.3 CURING REQUIREMENTS

Curing of the geopolymer-based concrete is different from that of conventional curing of the members in water. Preliminary pilot study tests revealed that fly ash-based geopolymer concrete did not harden immediately at room temperature (ambient curing). When the room temperature was less than 30°C, the hardening did not occur at least for 24 hours. Also, the initial mixing time is a very important parameter for GPC. The geopolymer concrete specimens should be wrapped during curing at elevated temperatures in a dry environment (in the oven) to prevent excessive evaporation. In trial mix, the curing temperature was 100°C and placed in the oven for 24 hours. Depending upon the test specimen sizes, the oven curing may be transformed to steam curing for the specimens, namely beams.

6.4 FRESH PROPERTIES

This section explains the primary responses of the fresh mixes of GPC for workability and compaction criteria. The tests for workability and compacting factors were conducted for GPC mixes with and without plastic waste inclusion. The readers are advised to refer the reference data and should execute the trial mixes with hands-on experience on the GPC concrete first.

6.4.1 WORKABILITY

6.4.1.1 Slump

Slump values indicate the stiffness of the fresh concrete mix. A specific range of acceptable slump values are given by the different national and international codes. The ranges of the slump values are indicated according to the application of the concrete for the foundation to structural members including column, slabs, beams, etc. The slump is a height measured in millimeters of the fresh mix measured with the help of a metal cone apparatus. The values discussed here are obtained from the experimental work carried out in accordance with IS: 1199-1959 code provisions. The internal surface of the mold is thoroughly cleaned and a light coat of oil is applied. The mold is placed on a horizontal, smooth, rigid, and nonabsorbent surface. With the freshly mixed concrete, the mold is then filled in four layers, each approximately one-fourth of the height of the mold. Each layer is tamped 25 times by the rounded end of the tamping rod. After the top layer is rodded, the concrete is struck off the level with a trowel. The mold is removed from the concrete by raising it slowly in the vertical direction. The slump value is obtained from the difference in the height of the cone and the height of the concrete cone after the slump.

Compared to cement-based concrete, the geopolymer concrete demonstrates stiffer mixture and shows less values of slump. This is due to the ingredients and their nature of formation. It is to be noted that the cement concrete does not require any additional time for the formation of the cement gel as the cement readily develops as a paste at the time of addition of water in the mix. The process of hydration takes place in a fraction of minutes. Therefore, during the wet mixing of the materials, the cement concrete acquires sufficient slurry form and is ready for use. The fresh mix of cement concrete is less viscous in nature. On the other hand, the waste-based concrete or the geopolymer concrete requires time to develop intermolecular bonding and the process of polymerization requires more time than the normal concrete. As a result, GPC mixes are more viscous in nature and differ in the behavior from the conventional one. The GPC mixes remain plastic in nature for a very less time as the alkaline solution is quick setting due to the gel paste developed by the mixing of fly ash and alkaline solution. The average slump values of the GPC mixes are less than that of normal concrete.

One more important aspect of GPC in fresh state is that the workable time of the GPC mixture is limited and requires quick actions for molding. The form works should be kept ready, and the pouring of concrete should be completed within 2–3 minutes of mixing. This is due to the nature of the alkaline chemical to set quickly. The author has observed flash setting of the mixes many times in case of GPC. Therefore, the use of appropriate admixture is necessary or rather unavoidable in GPC. From the experiments the average slump values of GPC mixes are observed to be between 50 and 80 mm. However, the values must be cross-checked by the experimental work and the molar content and alkaline binder ratio are noted to be influential parameters for the slump value behavior.

6.4.1.2 Compacting Factor and Flow Characteristics of the GPC Mixes

The compacting factor test indicates the ability of a fresh concrete mix for compaction ability of the mixture. It is obtained as the ratio of partial to fully compacted fresh mixtures. Along with the compacting factor, the flowability is also an important parameter of the workability of the concrete. The fresh mix is allowed to spread on its own for the given time and within the measured time, the mix should acquire the diameter necessary for the uninterrupted flow of the material. This behavior may be measured by a rotating or flow table apparatus. The values are indicated in percentage or as an index for the flow time for the given mix.

The compaction and flowability of concrete primarily depend on air voids and water content in the mix. If the material consists of dense formation, the compaction values can reach nearest to 1. Similarly, the flow index may reach up to the values 80 or 90. Following are significant parameters impacting the flow characteristics of concrete:

- Water content in the mixture
- Size of aggregates and availability of all-in-one aggregates
- Cement content in the unit volume of the mixture
- Use of admixture as accelerator
- Mixing time of constituents

Unlike conventional concrete, the GPC shows a slightly different behavior in the fresh state. The compaction factor values are reportedly higher than that of normal concrete. This is due to the fast processing of fly ash and alkaline liquid. The GPC mixes have been found to be more compaction friendly. The air voids are also less developed in the fresh state in GPC mixes. However, drying and settlement of the GPC concrete also occur at the faster rate, and at this stage the flash setting may take place; therefore, the placing of GPC should be done as early as possible after wet mixing is completed.

The flow characteristics of the GPC mixes are quite different than the conventional one. The alkaline solution is highly viscous in nature and the constituents namely sand and aggregates are very well lubricated in the very initial stage of wet mixing. This reduces the internal friction of the constituents. Moreover, the use of admixture namely accelerator continues to provide flexibility to the fresh mix for free flow conditions for some time. During this phase the material is highly flowable and shows good degree of dispersion in the cross sections of the moldings. Unlike the cement concrete mixes, the GPC mixes easily attain spread flow diameter of around 250–300 mm without any delay. The continuous hydration process of cement with water in conventional mixes hinders this behavior and increases the intermolecular friction of the constituents.

The overall workability of the GPC or the concrete prepared with wastes is one of the most sensitive properties and requires attention. Care should be taken regarding the time or delay in the process of dry and wet mixing stages. The GPC requires quick actions of mixing and placing processes compared to the cement concrete. The workability of GPC is observed to be governed by the ratio of alkaline liquid to fly ash and percentage of admixture in the mix.

6.5   MECHANICAL PROPERTIES

On observing the better density of waste-based GPC concrete, there can be a prediction possible that the mechanical strength properties should exhibit better performance than that of cement-based concrete. The denser matrix of the GPC concrete shows improved strength characteristics. Here in this section, the mechanical strength, namely compressive strength, splitting tensile strength, and pull-off strength test, results and effects of various concrete parameters on the response of the GPC have been discussed for better understanding of the influence of industrial wastes and plastic waste on the preliminary properties of the concrete. As far as the strength properties of GPC are concerned the most significant parameters are the molar content of the sodium hydroxide and the ratio of the liquid to fly ash portions. In the following subsections, the mechanical strength properties of the GPC mixes have been discussed to understand the influence of the varying fraction of molarity of the alkali solution and the liquid to powder ratio.

6.5.1 COMPRESSIVE STRENGTH

The compression test may be performed on cubes or cylinders. The cube specimen is of size $150 \times 150 \times 150$ mm; if the largest nominal size of the aggregate does not exceed 20 mm, 100 mm size cubes may also be used as an alternative. The cylinders possess length equal to twice the diameter. They are 150 mm in diameter and 300 mm long. Smallest test specimens may be used but a ratio of the diameter of the specimen to maximum size of the aggregate, not less than 3 to 1, is maintained. The compressive strength indicates the resistance capacity of the hardened material to the axially applied compression force. Compressive strength of concrete is measured by engineers in designing structures. The mean compressive strength required at a specific age is usually 7 days. Factors affecting the strength of concrete are ratio of SiO_2/Na_2O, $Na_2SiO_3/NaOH$, molarity of NaOH, degree of compaction, and curing temperature. The size of concrete cube used was $150 \times 150 \times 150$ mm. Compressive strength test on hardened fly ash-based geopolymer concrete was performed on a 2,000 kN capacity compression testing machine.

For reference and for general discussion, few of the results on the compressive strength tests have been discussed here. From Figures 6.1 and 6.2, the effect of varying molar content of NaOH and solution to binder ratio have been presented.

Effect of NaOH: As the concentration of solid sodium hydroxide increases in sodium hydroxide solution, compressive strength of geopolymer concrete increased for constant sodium silicate to sodium hydroxide ratio of 2 and 3 and alkaline liquid to fly ash ratio of 0.45 and 0.55. When NaOH concentration increases from 8 to 16 M, the compressive strength increased by 9%–45%.

Effect of $Na_2SiO_3/NaOH$: For sodium silicate to sodium hydroxide ratio of 2 and 3, compressive strength increased for ratio 3 than 2, by 1%–35% for constant alkaline liquid to fly ash ratio of 0.45 and 0.55 with NaOH concentration 8, 12, and 16 M.

Effect of alkaline liquid to fly ash: For the alkaline liquid to fly ash ratio of 0.45 and 0.55, compressive strength increased for 0.45 than 0.55 by 8%–30% for constant sodium silicate to sodium hydroxide ratio of 2 and 3 with constant sodium hydroxide concentration of 8, 12, and 16 M. From the graph, we observed that geopolymer concrete gave better result of compression strength for NaOH concentration of 12 and 16 M for the M25 grade concrete.

Alkaline/Fly ash = 0.45

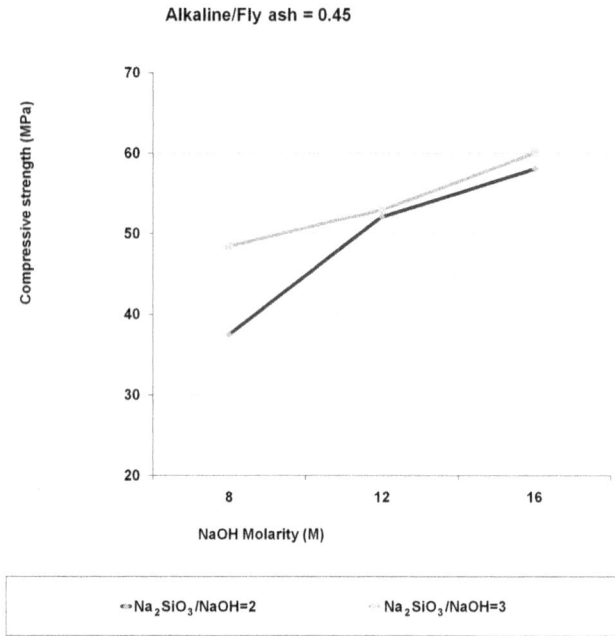

(a)

Alkaline/Fly ash = 0.55

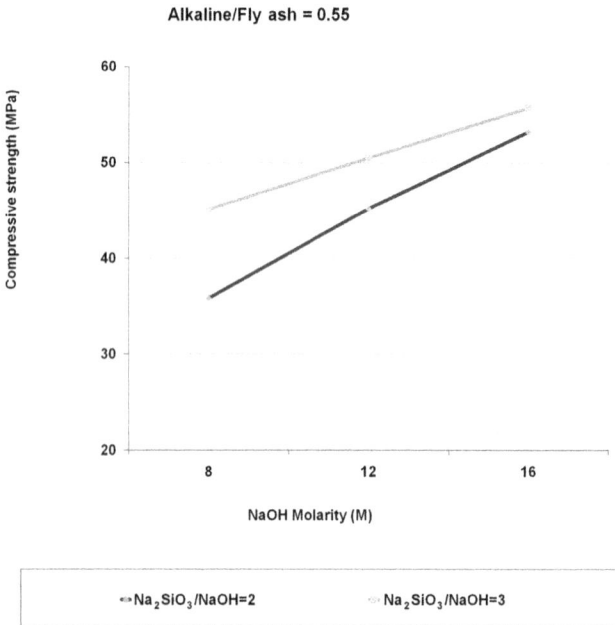

(b)

FIGURE 6.1 Effect of liquid to fly ash ratio on strength of GPC specimens. (a) Alkaline to fly ash ratio 0.45. (b) Alkaline to fly ash ratio 0.55.

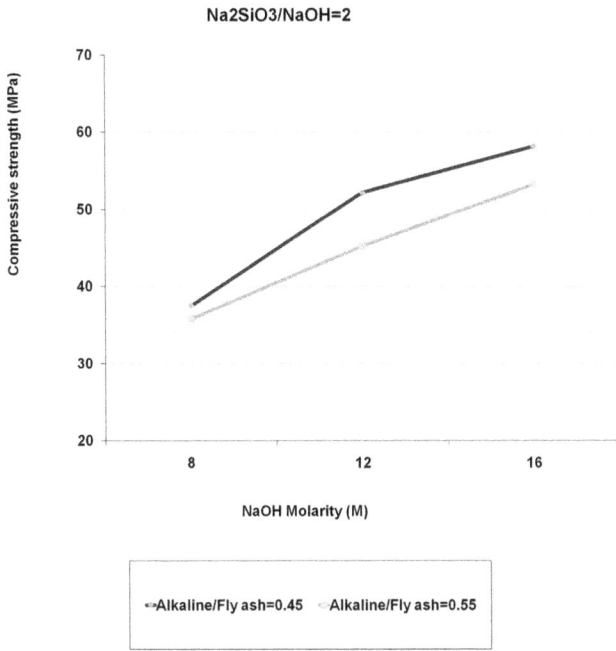

(a)

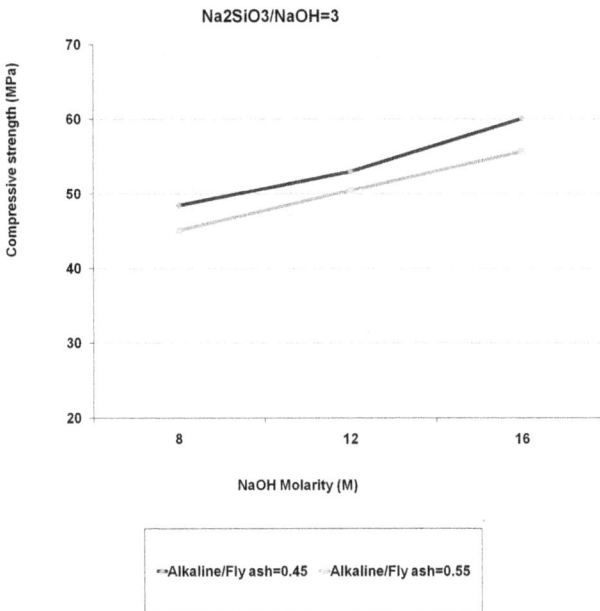

(b)

FIGURE 6.2   Effect of silicate to hydroxide ratio on strength of GPC specimens. (a) Sodium silicate to hydroxide ratio 2. (b) Sodium silicate to hydroxide ratio 3.

6.5.2 SPLITTING TENSILE STRENGTH

Direct tension tests of concrete are seldom carried out, mainly because the specimen holding devices introduces secondary stresses that cannot be ignored. The most used tests for estimating the tensile strength of concrete are the splitting tension test and the third-point flexural loading test. In the splitting tension test, a 150×300 mm concrete cylinder is subjected to compression loads along two axial lines that are diametrically opposite. The load is applied continuously at a constant rate within the splitting tension stress until the specimen fails. The compressive stress produces a transverse tensile stress, which is uniform along the vertical diameter. Size of concrete cylinder is 150 mm diameter $\times$ 300 mm height. The split tensile strength test on hardened fly ash-based geopolymer concrete was performed on a 2,000 kN capacity compressive testing machine. Split tension test was performed on cylinder after 7 days of casting. Split tensile strength of concrete cylinder was measured in N/mm^2. The steel plates were placed at the top and bottom between the platens of testing machine and the cylinder. The cylinder was placed with its axis horizontal between two steel plates on the universal testing machine.

Compared to conventional concrete GPC concrete shows improved values of strength in splitting actions also. Like compression strength, the resistance offered by the GPC to the indirect tension induced in the concrete was better due to the higher intermolecular bonding of the material. The reason for this behavior is interesting. In the case of general cement hydration process, there is always a possibility of air voids. However, the GPC mixes are denser due to the alkaline solution, unlike water in the conventional concrete. Moreover, the cement hydration process takes time and slowly gains strength with time. The GPC on the other hand shows faster rate of chemical reactions of fly ash and alkaline binders. Therefore, the complete hydration or, in this case, the polymerization in the GPC forms better internal strength against cracking by enhancing the stability and adhesion of the constituents. To understand the development of the strength mechanism against splitting actions, following results are presented for discussion (Figure 6.3).

Effect of NaOH: As the concentration of sodium hydroxide solid increases in sodium hydroxide solution, split tensile strength of geopolymer concrete increased by 10%–50% for constant sodium silicate to sodium hydroxide ratio and alkaline liquid to fly ash ratio.

Effect of Na$_2$SiO$_3$/NaOH: For sodium silicate to sodium hydroxide ratio increase, split tensile strength increased for ratio 3 than 2 by 1%–30% for constant sodium hydroxide concentration of 8, 12, and 16 M and alkaline liquid to fly ash ratio of 0.45 and 0.55.

Effect of alkaline liquid to fly ash: For the alkaline liquid to fly ash ratio, 0.45–0.55, split tensile strength increased for 0.55 than 0.45 by 2%–20% for constant sodium silicate to sodium hydroxide ratio of 2 and 3 with constant sodium hydroxide concentration of 8, 12, and 16 M. From the graph we may observe that the geopolymer concrete gave better result of split tensile strength for NaOH concentration of 12 and 16 M than the M25 grade concrete.

Na2SiO3/NaOH = 2

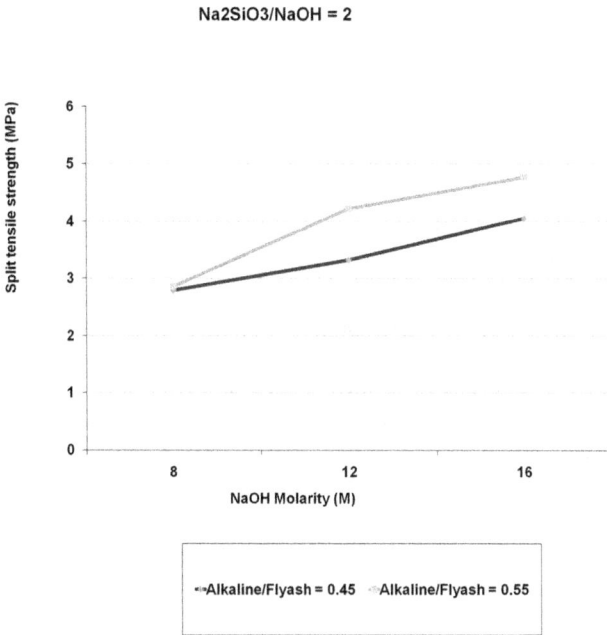

Alkaline/Flyash = 0.45   Alkaline/Flyash = 0.55

(a)

Na2SiO3/NaOH = 3

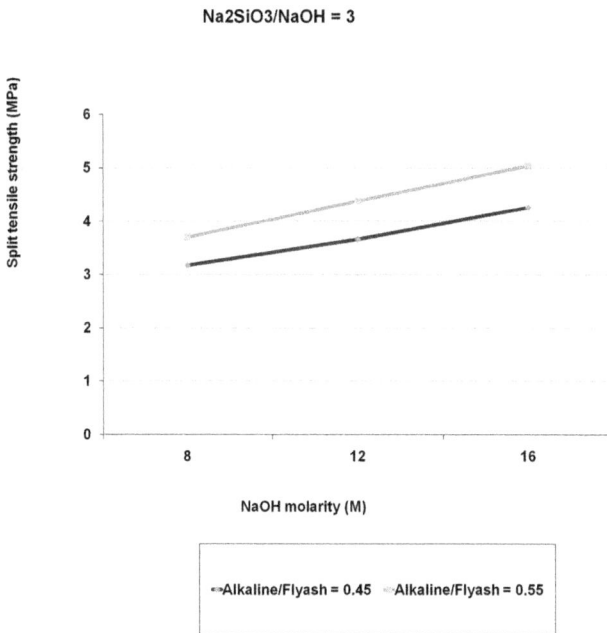

Alkaline/Flyash = 0.45   Alkaline/Flyash = 0.55

(b)

FIGURE 6.3 Effect of silicate to hydroxide ratio on splitting resistance of GPC. (a) Sodium silicate to hydroxide ratio 2. (b) Sodium silicate to hydroxide ratio 3.

6.5.3 IMPACT RESISTANCE

Sudden loading or falling load is one of the important load conditions in concrete structures. The concrete members subjected to machine loading or concentrated loading are especially susceptible for cracks and sudden failure. The capacity of concrete to withstand against the sudden loading without breaking or significant cracking is referred to as impact resistance. The specimens are evaluated for two stage conditions, namely the value of load generating initial cracking and the value of load at final cracking. Generally, the impact load effect is converted into the energy absorbed by the material using a standard relationship of weight, dropping weight distance, and number of blows or repeated loading on the samples. Therefore, it is also referred to as the dropping weight hammer testing.

The process of the testing is, however, simple; it requires meticulous observations on the samples facing the loads. The disk type of specimens is used and placed under the hammer mechanism. A metal sphere is placed at the center of the specimen and impacted with the free fall of the hammer. Each time when the hammer strikes the sphere, there is an impression created on the disk. After some repeated load blows, the specimen develops hairline cracks generally radially oriented. On continuous loading, the specimen develops a final failure. Both the initial and final cracks and number of blows corresponding to them are observed. The number of blows is converted into energy in joules and interpreted as the absorption capacity of the material.

In comparison, the specimens prepared with GPC mixes show better impact resistance to the cement concrete specimens. Better adhesion, highly dense material, and early strength of the hardened GPC matrix are the key aspects for better resistance to the impact. In case of conventional concrete, the intermolecular structure is not as dense as the GPC mixes and consists of voids and non-hydrated cement particles at the micro level. These limitations of hardened concrete restrict the effective stress distribution and increase the crack propagation within the external and internal mass of concrete.

In the following subsections, the impact resistance of the concrete by GPC specimens has been discussed. The tests on the disk specimens were conducted as per the suggestion of standard references and the apparatus was developed and fabricated at the laboratories by the author. The free fall of weight was managed manually, and the number of blows was counted and accordingly the cracks were referred and observed. The number of blows was convened in the energy absorbed by the disk specimen (Figures 6.4 and 6.5).

Effect of NaOH: As the concentration of sodium hydroxide solid increases in sodium hydroxide solution, impact resistance of geopolymer concrete increased by 10%–75% for constant sodium silicate to sodium hydroxide ratios 2 and 3 and alkaline liquid to fly ash ratios 0.45 and 0.55.

Effect of $Na_2SiO_3/NaOH$: For sodium silicate to sodium hydroxide ratios 2 and 3, impact resistance of geopolymer concrete increased for 3 than 2 by 5% and 20% for sodium hydroxide concentration 8, 12, and 16 M and alkaline liquid to fly ash ratio 0.45 and 0.55.

Na2SiO3/NaOH = 2, Alkaline/Flyash = 0.45

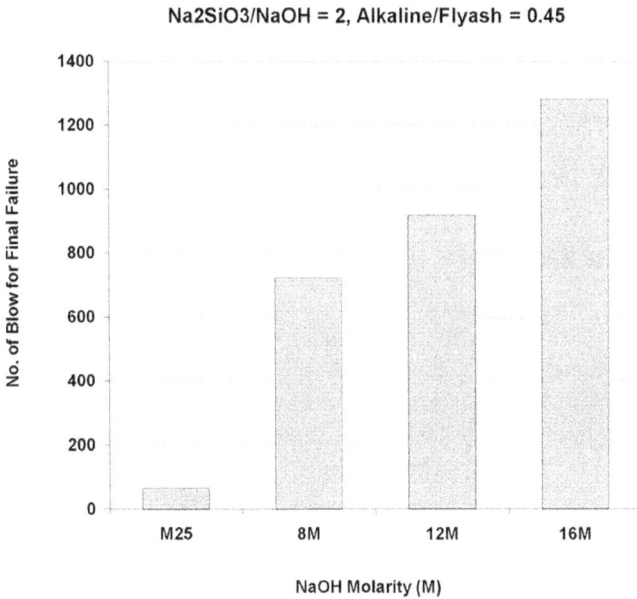

FIGURE 6.4 Effect of oxide ratios in alkaline solution on impact resistance of GPC with liquid to fly ash ratio 0.45.

Na2SiO3/NaOH = 2, Alkaline/Flyash = 0.55

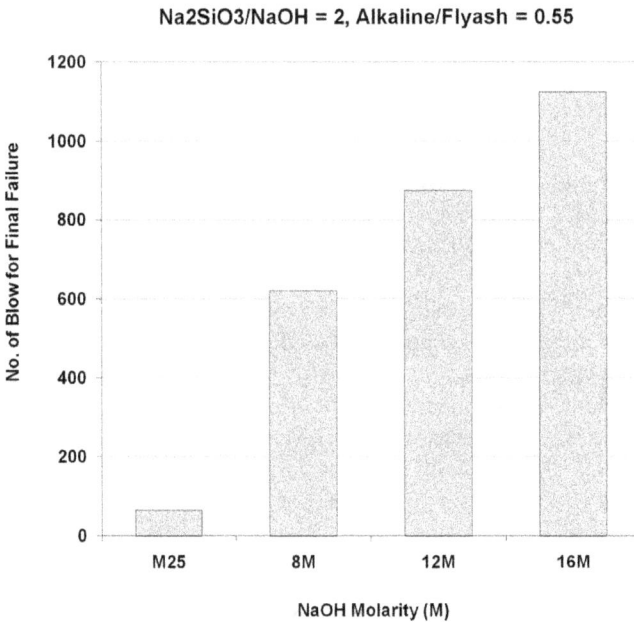

FIGURE 6.5 Effect of oxide ratios in alkaline solution on impact resistance of GPC with liquid to fly ash ratio 0.55.

Effect of alkaline liquid to fly ash: For the alkaline liquid to fly ash ratio 0.45 and 0.55, impact resistance of geopolymer concrete increased for 0.45 than 0.55 by 5%–25% for constant sodium silicate to sodium hydroxide ratio 2 and 3 with constant sodium hydroxide concentration of 8, 12, and 16 M. From the graph we observe that the geopolymer concrete gave much better result of impact resistance for NaOH concentration 8, 12, and 16 M than M25 grade concrete.

6.5.4 FLEXURE STRENGTH OF UNCONFINED GPC SPECIMENS

The results from the modulus of rupture test tend to overestimate the tensile strength of concrete by 50%–100%, mainly because the flexure formula assumes a linear stress–strain relationship in concrete throughout the cross section of the beam. Additionally, in direct tension tests the entire volume of the specimen is under applied stress, whereas in the flexure test only a small volume of concrete near the bottom of the specimen is subjected to high stresses; the flexure test is usually preferred for quality control of concrete for highway and airport pavements, where the concrete is loaded in bending rather than in axial tension.

The size of concrete beam specimen is 100 mm breadth $\times$ 100 mm depth $\times$ 500 mm length. Flexural bending test was performed on beam specimen after 28 days of casting. Flexural strength of concrete beam was measured in N/mm^2. The marking for support at 500 mm was done by pencil on beam specimen at three sides. The marking of two-point loads was also done by pencil on beam specimen at one-third distance from both support at three sides. During the flexural bending test on concrete beam specimen, it was observed that in concrete beam specimens cast from geopolymer concrete at failure load, the beam specimen breaks and separates into two parts. It was therefore observed that the preliminary response of the GPC specimens in flexure was found to be brittle but the load carrying capacity improved until the final failure.

This subsection discusses the influence of the chemical compositions of the alkaline solution and the molar contents of NaOH in the solution on the flexural strength response of the GPC specimens. The specific effect of the constituents, their concentration, and role in the matrix of GPC have been discussed in the following paragraphs.

Effect of NaOH: The molarity of sodium hydroxide (NaOH) solution increased as the flexural strength of geopolymer concrete increased. For a constant sodium silicate to sodium hydroxide ratio ($Na_2SiO_3/NaOH$), the flexural strength of geopolymer concrete increased with increase in the concentration of sodium hydroxide. The flexural strength was more prominent at 16 M of sodium hydroxide.

Effect of $Na_2SiO_3/NaOH$: As the ratio of sodium silicate to sodium hydroxide increased, the flexural strength also increased when $Na_2SiO_3/NaOH$ ratios were 1.0 and 2.0 in case SiO_2/Na_2O ratio was 2. However, the flexural strength decreased when $Na_2SiO_3/NaOH$ ratio was 2.0. While the silicate oxide to sodium oxide ratio was 3, the flexural strength increased throughout. For a constant molar concentration of NaOH, the flexural strength of geopolymer concrete increased with the increase in the ratio of $Na_2SiO_3/NaOH$. The flexural strength was more prominent at Na_2SiO_3

to NaOH ratio 2 when SiO_2 to Na_2O ratio was 2 and Na_2SiO_3 to NaOH ratio 3 when SiO_2 to Na_2O ratio was 3.

Effect of SiO_2/Na_2O: For a constant molar concentration of sodium hydroxide, the flexural strength of geopolymer concrete increased in silicate oxide to sodium oxide ratio 2 compared SiO_2 to Na_2O ratio 3. But when Na_2SiO_3 to NaOH ratio was 3, the flexural strength of geopolymer concrete increased in the SiO_2 to Na_2O ratio 3 compared to SiO_2/Na_2O ratio 2. The maximum flexural strength of geopolymer concrete for $Na_2SiO_3/NaOH$ ratio was 3 in case of SiO_2 to Na_2O ratio 3, while maximum flexural strength of geopolymer concrete for $Na_2SiO_3/NaOH$ was 2 in case of SiO_2 to Na_2O ratio 2. In general, the overall strength properties of GPC were found to be superior to conventional concrete.

6.6　DURABILITY PROPERTIES

Durability is referred as the resistance of the concrete to the various severe environmental conditions in general. For the laboratory investigations of the specimens prepared with the concrete of special types, the effects of the primary conditions, namely corrosion induced in reinforcements, external effects of acid and sulfate attacks, and most importantly the permeability or the ingress of water or air, are regarded as the parameters to evaluate the durability aspects. This section discusses the responses reported from the experimental program conducted on the specimens prepared with the GPC and plastic waste in various compositions.

6.6.1　Acid and Sulfate Attack

Acid attack is one of the most important aspects for consideration when we deal with the durability of concrete. Acid attack is particularly important because it primarily causes corrosion of reinforcement. Statistics have indicated that over 40% of failure of structures is due to corrosion of reinforcement. The total specimen of nine cubes was kept in normal room temperature for 7 days after casting. After 7 days, three cubes were kept in a dry place and the weight of dry cube specimen was taken. Then the dry specimen was immersed in acid curing storage tank for up to 28 days. After 28 days, the cubes were taken out and kept in a dry place. After removing the worst surface, again weight of specimen was taken and the weight difference and compressive strength of specimens were measured. The cubes were cast at size of $150 \times 150 \times 150$ mm and kept at a temperature of 100°C at oven for 24 hours. After 24 hours the cubes were removed from the mold and kept in normal room temperature. After 7 days the cubes were immersed in a 5% concentric sulfuric acid (H_2SO_4) and 5% hydrochloric acid (HCl). After 28 days of curing, the weight difference and the compressive strength of cubes were measured.

In a similar way, the sulfate resistance of the specimens was obtained. The test procedure for sulfate resistance test was developed by modifying the related standards for normal Portland cement and concrete (Standards ASTM, 1993, 1995, 1997; Standards-Australia, 1996b). Total nine cubes of specimen were kept in normal room temperature for 7 days after casting. After 7 days, of nine cubes three cubes were taken and kept in a dry place and the weight of dry cube specimen was measured.

The specimen was then immersed in sulfate curing tank for up to 28 days. After 28 days, the cubes were taken out and kept in a dry place and the weight of specimen along with the weight difference and compressive strength of specimens were measured. The cubes were cast at a size of $150 \times 150 \times 150$ mm and kept at a temperature of 100°C at oven for 24 hours. After 24 hours the cubes were removed from the mold and kept in normal room temperature. After 7 days curing the cubes were immersed in a 10% sodium sulfate solution. After 28 days of curing, the weight difference and compressive strength of cubes were measured. The results of the tests have been presented in the form of a graph for understanding the response of GPC with the quantification of the resistance.

Effect of NaOH: When the molarity of sodium hydroxide (NaOH) solution increases, results show that the weight in acid attack and sulfate attack is reduced. In acid curing, the molarity of NaOH increased then reduced for the weight at 0.3%–0.7%. The test results indicate that weight decreased at 0.3%–0.8% in acid curing compared to sulfate curing. In acid curing, the weight of specimens increased from 3% to 4% and in sulfate curing the weight increased from 4% to 5% in normal condition.

Effect of Na_2SiO_3/NaOH: When the sodium silicate to sodium hydroxide ratio was 2, the weight was reduced compared to other sodium silicate to sodium hydroxide ratios. The weight of geopolymer concrete increased with the increase of sodium silicate to sodium hydroxide ratio when the silicate oxide to sodium hydroxide ratio was 2. But the sodium silicate to sodium hydroxide ratio increased when the weight of geopolymer concrete decreased for 12 M and 14 M of NaOH in case of silicate oxide to sodium oxide ratio of 2. The weight of geopolymer concrete was more prominent at Na_2SiO_3 to NaOH ratio of 1 in both curing conditions.

Effect of SiO_2/Na_2O: For all molar concentrations of sodium hydroxide, the weight of geopolymer concrete increased in silicate oxide to sodium oxide ratio 2 compared to SiO_2 to Na_2O ratio 3 for all Na_2SiO_3 to NaOH ratios. Weight of geopolymer concrete was more prominent at SiO_2 to Na_2O ratio 3 in acid attack and sulfate attack.

6.6.2 Oxygen Permeability of GPC

The fluid ingress into a material through the surface with or without the external pressure conditions can be referred as the permeability. In case of concrete, it is regarded as the permeability coefficient expressed as the rate of passing of the liquid or gas through the surface of the specimens in a controlled environment. The cross section of the specimen is examined for its ability to pass through the fluid in the given time interval and the pressure exerted by the fluid. Based on the experimental results following graphs are shared for more results and interpretations. The source of fluid was oxygen gas supplied at a constant pressure head.

Effect of NaOH: Permeability of geopolymer concrete decreased with the increase of molarity of sodium hydroxide (NaOH) solution in case of Na_2SiO_3/NaOH ratio of 2 and 3. While the sodium silicate to sodium hydroxide ratio (Na_2SiO_3/NaOH) was 1, the permeability of geopolymer concrete was increased with increase in the concentration of sodium hydroxide. The permeability of geopolymer concrete was more prominent at 8 M of sodium hydroxide in a SiO_2 to Na_2O ratio of 2.0.

Effect of Na_2SiO_3/NaOH: The permeability of geopolymer concrete increased with the increase of sodium silicate to sodium hydroxide ratio. But the sodium silicate to sodium hydroxide ratio increased when the permeability of geopolymer concrete decreased for 16 M of NaOH in case of silicate oxide to sodium oxide ratio of 2. While the silicate oxide to sodium oxide ratio was 3, the permeability of geopolymer concrete increased throughout. The permeability of geopolymer concrete was more prominent at a Na_2SiO_3 to NaOH ratio of 1.

Effect of SiO_2/Na_2O: For a constant molar concentration of sodium hydroxide, the permeability of geopolymer concrete increased in silicate oxide to sodium oxide ratio of 3 compared to SiO_2 to Na_2O ratio of 2 in case Na_2SiO_3 to NaOH ratios were 2 and 3. While the Na_2SiO_3 to NaOH ratio was 1, the permeability of geopolymer concrete decreased in silicate oxide to sodium oxide ratio 3 compared to SiO_2 to Na_2O of 2. Permeability of geopolymer concrete was more prominent at SiO_2 to Na_2O ratio 3.

Besides the common observations and analysis of the above results, there are a few salient features observed from the tests conducted on the specimens. The important observation was that the permeability is found to be the function of the overall air void ratio of the material. In this regard, the GPC shows better intermolecular adhesion as earlier discussed. Also, the GPC concrete develops strength at an early age and does not show water absorption for a given condition where the specimen is under controlled and nearly vacuumed conditions.

6.6.3 WATER ABSORPTION

Water absorption test was used to measure the capillary water absorption of concrete. Some materials with extremely coarse pore structure experience little capillary suction and may show significant deviation from linearity after prolonged wetting. Capillary suction can be measured in dry concrete. Sorption does not take place in saturated materials, and in totally dry materials substantial absorption of water by the gel will distort the results. The absorption will depend on the initial water content and its uniformity throughout the specimen under test. Furthermore, as water absorption and capillary suction depend on porosity, any non-uniformity in the latter could lead to different absorption in samples obtained from what is supposed to be the same material. It is, therefore, essential that materials under test be consistent and homogeneous. Water absorption test is performed on same disk used for oxygen permeability test. Oxygen permeability and water absorption of concrete disk were measured in the same direction. During the water absorption test on concrete disk, it was observed that the water absorption at 1,800 seconds was higher than at 300 seconds in cast disk. For details, sample results are shown and discussed regarding the performance of the GPC in resistance to water absorption. While performing the tests care should be taken as the sides of the concrete cylinder disk were sealed by white grease to avoid evaporative effect as well as to maintain uniaxial water flow and the opposite faces were left open. Before the testing, their initial weights should be recorded. The concrete disk was submerged 3–5 mm in water. The water absorption at 300- and 1,800-second intervals should be measured with scale of 0.1 g readability.

Effect of NaOH: The water absorption of concrete cast disk decreased with the decrease in molar content of sodium hydroxide (NaOH) solution, but at 14 M of sodium hydroxide, water absorption at 300 and 1,800 seconds increased beyond that for 12 M and 16 M. The water absorption at 300 and 1,800 seconds was more prominent at 8 M of sodium hydroxide.

Effect of Na_2SiO_3/NaOH: The water absorption of geopolymer concrete increased with the increase in sodium silicate to sodium hydroxide ratio when silicate oxide to sodium oxide ratio was 2 compared to 3. But the sodium silicate to sodium hydroxide ratio was increased; the water absorption of geopolymer concrete decreased for 8 M and 16 M of NaOH in case of silicate oxide to sodium oxide ratio of 3. While the silicate oxide to sodium oxide ratio was 2, the water absorption of geopolymer concrete increased throughout. The permeability of geopolymer concrete was more prominent at a Na_2SiO_3 to NaOH ratio of 3 in case SiO_2 to Na_2O ratio was 3.

Effect of SiO_2/Na_2O: For a constant molar concentration of sodium hydroxide, the water absorption at 300 and 1,800 seconds of geopolymer concrete decreased in silicate oxide to sodium oxide ratio 3 when compared with SiO_2 to Na_2O ratio 2 in case Na_2SiO_3 to NaOH ratios were 2 and 3. Water absorption at 300 and 1,800 seconds of geopolymer concrete was more prominent at SiO_2 to Na_2O ratio 3.

6.7 STRUCTURAL PERFORMANCE OF MEMBERS PREPARED WITH MODIFIED CONCRETE

6.7.1 FLEXURE BEHAVIOR

The GPC members consisting of PCMPW were assessed for performance under the structural loading conditions, namely in compressive flexure actions. The study of the structural performance of a new or modified material is always of great importance. The beams were prepared with the varying amount of the reinforcements. The aim of the tests was to identify the influence of the plastic wastes on the conventional response of the beams. Moreover, the waste-based concrete, i.e., the GPC, was also to be studied with the normal concrete. However, the reinforcements were used in the beam specimens, and the unreinforced beam members were also prepared and tested in flexure.

The load to deflection results show that the addition of the plastic waste in GPC influences the load carrying capacity of the beams. Though the failure pattern and the responses of the members were governed by the reinforcement as a typical behavior, the failure mode of the concrete due to better adhesion of the constituent showed differences to a great extent. Especially in the compression zone, as shown in Figure 6.6, the crushing of the concrete was observed to be limited, which otherwise exhibits more scattered and brittle failure of the material in the top fiber zone. While in the case of GPC consisting of denser concrete mass, less brittle failure of the material was seen.

Along with the improved material behavior, the GPC showed comparatively better load carrying capacity with and without plastic wastes. The test results are discussed here for reference. The load to deflection relation of normal or conventional concrete as well as GPC concrete consisting of the waste from the industry, without the use of cement, has been demonstrated with the graph in Figures 6.7 and 6.8.

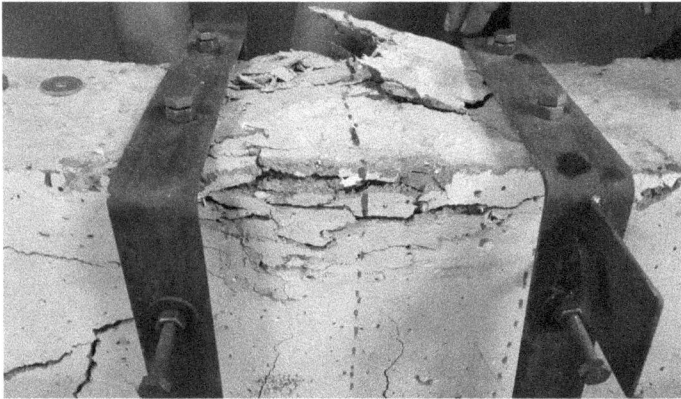

FIGURE 6.6   Limited failures in the compression zone of GPC concrete.

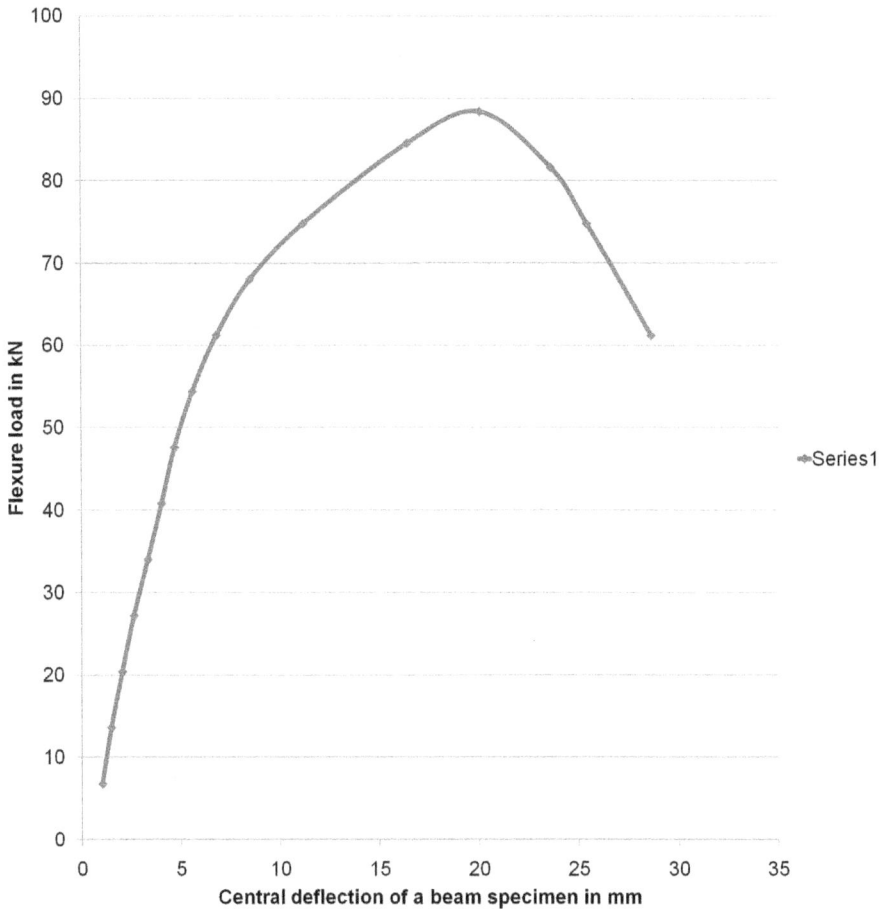

FIGURE 6.7   Load to deflection relationship of conventional concrete beam.

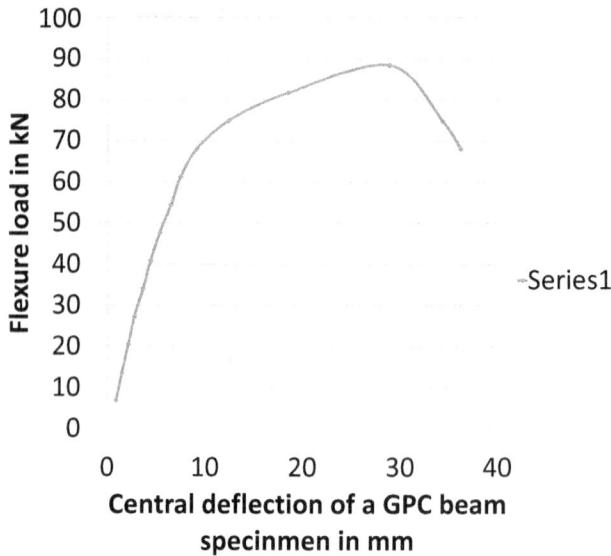

FIGURE 6.8 Load to deflection relationship of GPC beam.

By direct comparison of the results, it may be observed that the GPC beam showed larger value of the deflection for the given loading. This ensures better adhesion capacity of the material and strength to resist the forces before failure. It is to be noted that the reinforcement must play its own role of resisting the tension induced into the beams while loading; however, the compression zone is more susceptible for failure by crushing. The load carrying capacity of a flexure member primarily depends on the tensile capacity of the material as well as loss of material in the compression zone by crushing. The beams prepared with GPC showed better and improved load carrying capacity compared to the normal concrete even with the presence of the reinforcement.

6.7.2 AXIAL STRESS–STRAIN BEHAVIOR OF THE CONCRETE SUBJECTED TO COMPRESSION

One of the important properties of concrete is ductility manifested by the elastic modulus and stress–strain relationship. The stress–strain relationship is important evidence to understand the failure pattern of the material under and above peak stress values and how the corresponding strain is being developed in the material before the final failure. The typical response of the concrete in axial compression for the elastic properties is largely brittle beyond the peak stress values. Moreover, the collapse of the material is sudden and shows inelastic response with a limited strain value

TABLE 6.4

Elastic Properties of GPC and Influence of Molar Content

	NaOH (M)	Compressive Strength (MPa)	Ec (GPa)	Strain at Peak Stress	Poisson's Ratio
$Na_2SiO_3/NaOH = 2$	8 M	27.81	12.64	0.0024	0.12
Alkaline/fly ash = 0.45	12 M	35.07	15.25	0.0036	0.13
	16 M	42.95	24.54	0.0041	0.17
$Na_2SiO_3/NaOH = 2$	8 M	26.5	11.6	0.0033	0.13
Alkaline/fly ash = 0.55	12 M	33.43	15.22	0.0025	0.12
	16 M	38.99	17.82	0.0026	0.15
$Na_2SiO_3/NaOH = 3$	8 M	37.27	16.56	0.0026	0.15
Alkaline/fly ash = 0.45	12 M	41.66	21.64	0.0031	0.17
	16 M	45.1	25.77	0.0042	0.16
$Na_2SiO_3/NaOH = 3$	8 M	33.65	15.13	0.0025	0.13
Alkaline/fly ash = 0.55	12 M	37.42	17.09	0.0041	0.14
	16 M	44.68	25.16	0.0033	0.17

of 0.0002. In this section, the behavior of the cylinder specimens subjected to axial forces for the determination of the stress and strain at peak stress and strain at final failure is discussed in brief. In Chapter 5, the stress–strain relationship of concrete consisting of PCMPW was discussed in detail. With reference to that behavior, the stress–strain performance of the GPC has been discussed here.

It is already mentioned that GPC exhibits better strength compared to normal concrete in nearly all test conditions, and hence, the same behavior may be observed in the case of stress–strain relationship also. Table 6.4 shows the elastic values of GPC concrete along with the stress–strain values and strain at peak stress. The table also shows the influence of the molar content on the properties of the GPC. The results are the actual experimental data observed during some of the investigations carried out by the author and his students.

The test data in the form of graphs are shown in Figures 6.9 and 6.10 for reference. It may be observed that the stress–strain relationship is more toward the linear performance of the material before and after peak stress values, and that is exactly where the GPC differs from normal concrete.

As indicated from the figures, the GPC specimens showed increased values of the failure strain compared to the typical strain range of the conventional concrete between 0.0002 and 0.00035. It is to be noticed that failure strain value of any concrete like composite depends on the intermolecular adhesion, crack development at the given loading values, crack propagation within the matrix, and crushing behavior of the material at extreme loading conditions. In all such parameters, the GPC showed far better and improved results compared to conventional cement-based concrete. Needless to add, more the failure strain value, better is the structural behavior

Na₂Sio₃ / NaOH=2,
Alklaine/Fly ash=0.45,NaOH=8M

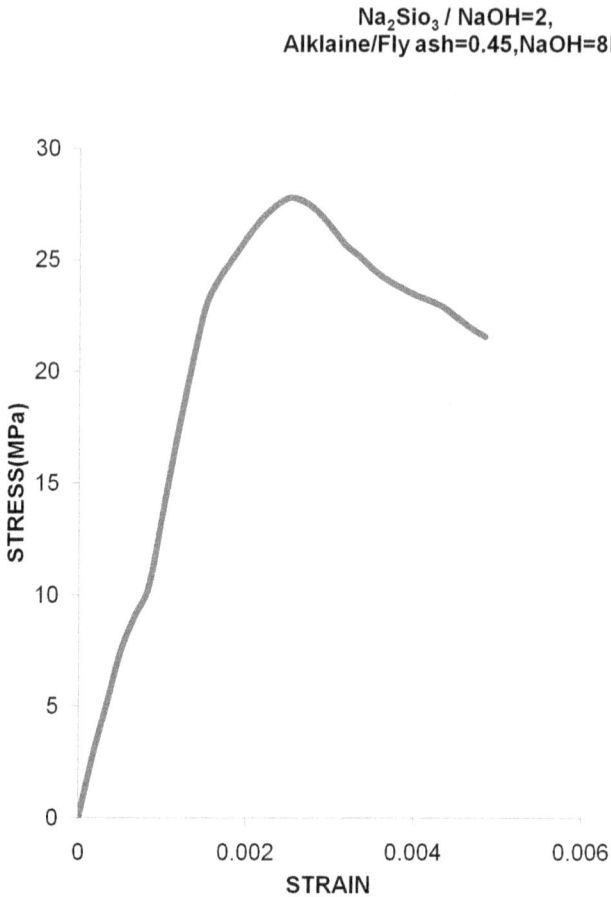

FIGURE 6.9 Stress–strain relationship of GPC subjected to axial force of 8 M NaOH concentration.

of the material in the axial loading conditions. The graph also indicates important influence of the molar content of sodium hydroxide in the alkali activator solution. Higher the molar content, better the strength of the mixture. This is helpful to indicate that the mix design should be prepared in a way that the desired strength or target strength of the concrete may be achieved by adjusting the molar content of NaOH in the liquid solution.

Na$_2$Sio$_3$ / NaOH=2,
Alkaline/Fly ash=0.45,NaOH=12M

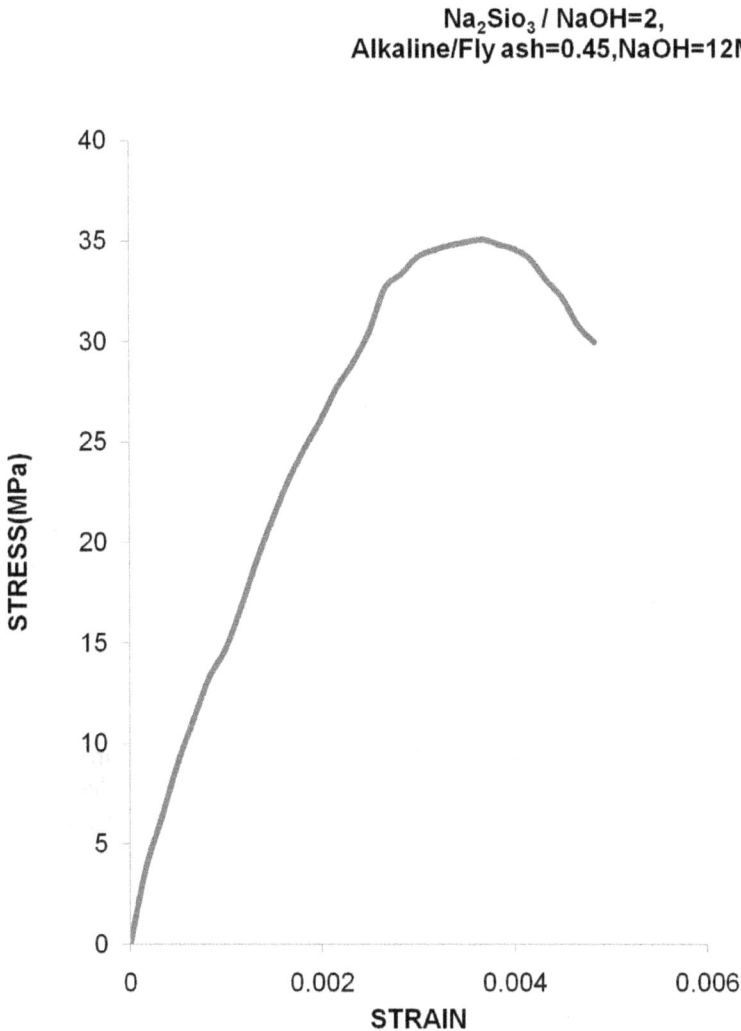

FIGURE 6.10 Stress–strain relationship of GPC subjected to axial force of 12 M NaOH concentration.

6.8 SPECIAL REMARKS ON THE USE OF DIFFERENT INDUSTRIAL WASTES IN CONCRETE

6.8.1 FLY ASH

In this section, fly ash being one of the most commonly available industrial wastes, its use is being discussed and demonstrated. The section shows how fly ash may be used instead of cement to form the binder material for concrete preparation. It is to be noted that all mixes have been prepared and tested using class F fly ash only.

This is because the GPC required polymerization to take place when the dry powder binder materials are mixed with the liquid solutions. The class F fly ash consists of silicon dioxide as the main constituent. The alkali activators need silica-rich materials to form the multilayered structural bonds with sodium content. However, without going into the detail of the chemistry of the GPC, the role of a specific mineral is important. The class F fly ash is generally obtained from the thermal power stations where anthracite-based coal is used as fuel. On the other hand, the class C type of fly ash may be obtained from the power stations wherein lignite-based coal is used as fuel.

The type of fly ash should be clearly identified, rather should be investigated because there are important points of observations to discuss. Class F fly ash releases plentiful silica contents in the matrix to generate bonds with alkali activators. On the other hand, class C fly ash consists of calcium oxide as its major content. This means class C fly ash possesses more energy to develop intermolecular adhesion with the other concrete making constituents. Moreover, compared to class F fly ash, class C fly ash holds better binding capacity. However, geopolymerization is an exothermic process and, therefore, addition of excessive calcium results in cracking of the hydrated binder material paste. More shrinkage and post-plastic shrinkage cracks are expected in the hardened mass also.

6.8.2 Marble Wastes

The cutting and finishing processes of the raw marble stones generate huge amount of waste. The wastes are in the form of sludge and sand. The sands are also found in the fine, finer, and micro fine particle sizes. The sludge, on the other hand, possesses primarily the powder in a slurry form. On drying or dewatering of the sludge, the fine powder of the marble may be obtained. Following points should be remembered while using marble wastes in making of concrete and mortars.

Particle size distribution tests must be employed on the samples. The replacement of the natural sand should be carried out according to the particle size of the natural sand. The slurry or sludge form of the waste should be checked for maximum replacement or the upper limit should be specified in case the cement is to be replaced. Preliminary tests on the waste sample conforming to the relevant IS codes must be employed and the feasibility of the addition of the sand should be confirmed by result analysis. The grain sizes of the marble sand are more uniform compared to natural sand. Therefore, there are possibilities of more air voids in the mixture. Though this will support better workability of the mixture at the initial stage, the hardened mass may struggle in obtaining the required strength under loading actions. If the sludge is being used, the water to cement ratio of the mortar or concrete should be balanced adequately for the workability of the mixture. The sludge in dry form needs more water since it possesses water retention capacity unlike the cement powder. The sludge does not exhibit any binder characteristics. In this case, when the cement is replaced with the sludge there should be adequate use of chemical admixture for better results in the adhesion mechanism of the constituents.

In general, marble wastes are one of the promising wastes that may be used as an alternative to natural materials, however, with sufficient trials and strength confirmation.

6.8.3 USED FOUNDRY SAND

Steel industry involved in the casting process generates a special type of waste referred to as foundry sand. The casting process uses molds prepared with the chemically bonded sand. The sand is finer than the regular natural sand and possesses all typical properties of natural sand. While utilizing the used foundry sand in the concrete, following points are to be taken care. The foundry process requires to bond the sand by admixture and chemicals. Therefore, the waste sand should be checked for chemical composition before use. The foundry sand may be the best replacement of natural sand. However, the physical properties are like the natural sand, and so pilot tests on the sand samples from the waste should be performed. Unlike natural sand, the waste foundry sand contributes to the intermolecular bonding of the matrix. Therefore, it is well expected that the concrete or mortar may perform better in the application to the special structures. Including all preliminary properties, density is one of the important parameters for materials and therefore, the density and specific gravity along with the water absorption of the waste foundry sand should be cross-checked adequately. The waste foundry sand has been observed to be contributing to better compressive strength response. Along with increased resistance to the splitting actions on the specimens, the waste shows good agreement with improved crack resistance also. As far as the durability of concrete or mortar is concerned, the waste foundry sand increases the resistance against permeability. This is due to the finer size of the particles and from the partial bonding existing in the particles. However, the responses by the waste foundry sand are encouraging for utilization in the concrete, and adequate pre-treatments are necessary. The chemically bonded sand is largely discarded in lump form. Therefore, grinding or crushing processes are necessary for the waste material. Regarding the water absorption of the sand, the experimental investigations by the author and his team have observed that the overall rate of water absorption reduces with the replacement of the natural sand in the mortar up to 30% fraction by weight of the natural sand.

In this section, the development of concrete and mortars prepared with the industrial wastes was discussed elaborately by mentioning the mix design, test results, and the influence of various attributes of the GPC concrete on the workability, strength, and durability aspects. The structural response by the members prepared with waste-based concrete was also explained with the major outcomes of the experimental investigations. Finally, special notes and some important remarks on the use of other industrial wastes were discussed for knowledge and information for the readers.

7 Evaluation of the Concrete Blended with Industrial and Plastic Wastes

For this section, the experimental results presented for discussion are all based on the specific mix design and the size of the PCMPW fibers. For clarity, the mix design of the GPC utilized during the work and the general properties of the plastic fibers are mentioned in Table 7.1. The mixes were prepared for 24 hours at 100°C curing temperature in a hot air oven at the laboratory. The general properties of the PCMPW are: thickness – 60 µm, category – metallized food packing grade, type – polythene film (metallized), and density – 0.98 to 1.412 g/cm^3.

7.1 FRESH AND MECHANICAL PROPERTIES

The fresh property of the concrete includes workability and compacting ability of the material. The workability of the concrete consisting of plastic wastes and industrial wastes in case of geopolymer concrete is presented and discussed.

TABLE 7.1
Mix Proportions of the GPC Constituents with Trial Batch Results

Batches Parameters	Batch B1	Batch B2	Batch B3	Batch B4	Batch B5	Batch B6	Batch B7
Liq./FA	0.5	0.5	0.5	0.5	0.5	0.5	0.5
Na$_2$SiO$_3$/NaOH	1	1	1	1	1	1	1
SiO$_2$/Na$_2$O	2.25	2.25	2.25	2.25	2.25	2.25	2.25
Molarity of NaOH	8	8	8	8	12	12	12
Metallized plastic	0%	0.5%	1%	1.5%	0%	0%	0.5%
Super plasticizer	0.5%	0.5%	0.5%	0.5%	0.5%	0.5%	0.5%
Extra water	0%	10%	10%	10%	10%	15%	15%
Age of concrete	7 days	7 days	7 days	7 days	7 days	7 days	7 days
Density in kg/m^3	2,461	2,318	2,313	2,263	2,310	2,267	2,205
Comp. strength (MPa)	34.22	20.74	18.37	12.44	15.8	13.33	9.92
Splitting tensile strength (in MPa)	2.26	1.27	1.13	0.707	–	0.60	0.81

DOI: 10.1201/9781032621340-7

7.1.1 Fresh State Behavior and Properties

In general, the primary factors affecting the fresh behavior of concrete are:

- The water cement ratio
- Use of admixture in concrete to enhance the flow characteristics especially in case of self-compacting concrete
- Aggregate and particle sizes
- Cement and inter-material dry mixing
- Use of filler materials with pozzolanic nature
- Time and method of mixing the dry constituents and wet slurry
- Vibration induced by machine and manual methods
- Use of all-in-one aggregates or specific types and size of aggregates

Apart from the abovementioned factors, the concrete shows high sensitivity toward the addition of plastic. The earlier chapters described the response of the addition of plastic waste in concrete with the experimental results. The plastic influences the fresh state behavior of concrete.

7.1.2 Workability of GPC with PCMPW

7.1.2.1 Slump Test and Behavior

The slump test is used to gauge the characteristic of brand-new concrete. It is an empirical measurement that gauges how easily new concrete can be worked. Due to the straightforward apparatus and straightforward process, the test is well-liked. To guarantee uniformity for different batches of comparable concrete under field conditions, the slump test is used to assess the effects of plasticizers upon their introduction. A method of evaluating the uniformity of new concrete is the slump test. Slump tests allow for the observation of various slump varieties. The three distinct kinds of slump are collapse, shear, and true slump. The concrete entirely collapses during a collapse slump. However, this response reflects that the concrete is not useful for structural applications, namely casting of beams and column, unless it is made for use in self-compacting concrete.

During the experimental work, true slump of concrete is observed. In a concrete mix of $Na_2SiO_3/NaOH$ and SiO_2/Na_2O containing 8 to 16 M NaOH and 0.5% super plasticizer with metallized polymer plastic, the slump was in the range of 18–85 mm. For an exact understanding of the influence of chemical compositions of GPC and PCMPW on slump behavior, following graphs (Figures 7.1 and 7.2) may be referred and the discussion is presented for better learning on the topic.

From the sample test results it may be observed that the PCMPW fibers in the GPC reduce the flowability of the fresh mixes. Compared to conventional cement-based concrete, the GPC shows stiff behavior in the fresh state. Moreover, the addition of the plastic fibers further increases the friction with the constituents at the higher dosage of the plastic. The effect of molar content of NaOH and oxide ratios of the mix also influence the fresh behavior and response of the mixes. Following are the general observations retrieved from the results and presented for better understanding of how the GPC performs with plastic waste.

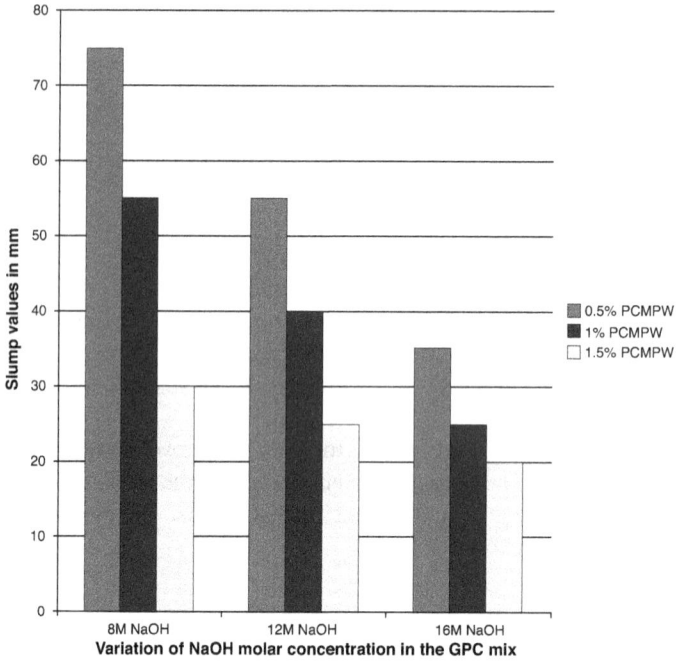

FIGURE 7.1   Effect of hydroxide ratio and plastic fiber fractions on the slump.

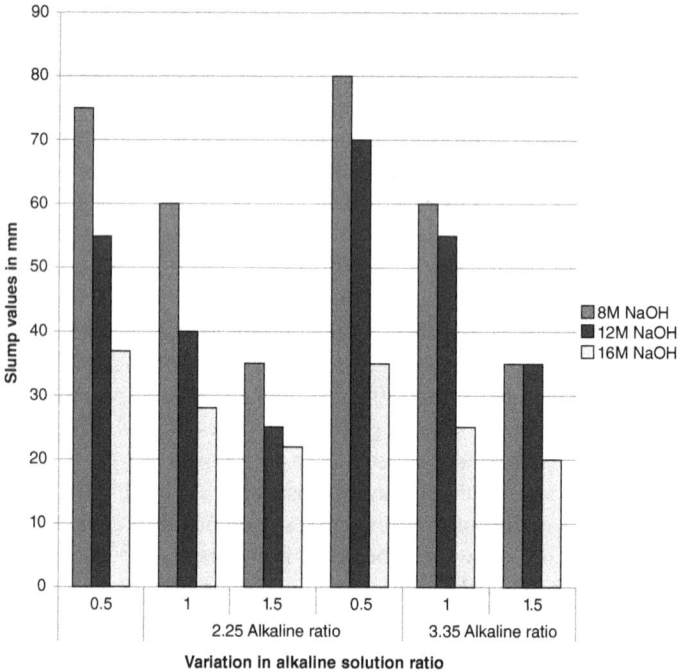

FIGURE 7.2   Effect of oxide ratio, molar content, and plastic fiber fractions on the slump.

Effect of NaOH: With increase in molarity of sodium hydroxide (NaOH) solution, the slump of concrete decreases. For a constant sodium silicate to sodium hydroxide ratio ($Na_2SiO_3/NaOH$), the slump of geopolymer concrete using metallized polymer plastic decreased with the increase in the concentration of sodium hydroxide.

When the NaOH was increased from 8 to 12 M, the slump decreased by 12.5%–26.66%, 8.33%–26.66%, and 16.66%–20% and when the NaOH was increased from 12 to 16 M, the slump decreased by 10.71%–36.36%, 9.09%–37.5%, and 10%–20% for the polymer plastic of 0.5%, 1%, and 1.5%, respectively, in case of SiO_2/Na_2O ratio of 2.25. When the NaOH was increased from 8 to 12 M, the slump decreased by 5.40%–20%, 6.67%–25%, and 12%–14.28% and when the NaOH was increased from 12 to 16 M, the slump decreased by 20%–38.33%, 21.42%–37.77%, and 9.09%–26.67% for the polymer plastic of 0.5%, 1%, and 1.5%, respectively, in case of SiO_2/Na_2O ratio of 3.35.

Effect of $Na_2SiO_3/NaOH$: As the ratio of sodium silicate to sodium hydroxide increases, the slump of the concrete decreases. For increase in $Na_2SiO_3/NaOH$ ratio from 1.0 to 2.0 the slump decreased by 6.25%–21.42%, 8.33%–27.27%, and 14.28%–28.57% and for increase in $Na_2SiO_3/NaOH$ ratio from 2.0 to 3.0 the slump decreased by 28.27%–53.33%, 20%–45.45%, and 10%–20% for the polymer plastic of 0.5%, 1%, and 1.5%, respectively, in case of SiO_2/Na_2O ratio of 2.25. When the $Na_2SiO_3/NaOH$ ratio increased from 1.0 to 2.0 the slump decreased by 11.76%–20%, 7.69%–25%, and 12.5%–25% and for increase in $Na_2SiO_3/NaOH$ ratio of 2.0–3.0 the slump decreased by 24.32%–50.66%, 21.42%–50%, and 9.09%–28.57% for the polymer plastic of 0.5%, 1% and 1.5%, respectively, in case of SiO_2/Na_2O ratio of 3.35.

Effect of metallized plastic fibers: For a constant molar concentration of sodium hydroxide and constant $Na_2SiO_3/NaOH$ ratio the slump of the concrete decreased with the increase of metallized polymer plastic. When metallized plastic waste increased from 0.5% to 1% and 0.5% to 1.5%, the reduction in the slump was in the range of 14.28%–28.57% and 28%–60% for SiO_2/Na_2O ratio of 2.25, and 18.91%–25% and 28.57%–53.33% for SiO_2/Na_2O ratio of 3.35.

7.1.2.2   Compaction Factor

The compaction factor indicates the ability of the fresh mix to the degree of compaction. The test requires measurement of the weight of the partially and fully compacted concrete and the ratio of partially compacted weight to the fully compacted weight, which is always less than 1, known as compaction factor. For the normal range of concrete the compacting factor lies between 0.80 and 0.92. This test is particularly useful for dryer mixes for which the slump test is not satisfactory. The compaction factor is generally unsatisfactory for values greater than 0.92. To obtain fair behavior, very low quantity of oil is applied to all the inside surfaces of hoppers to reduce friction. In compaction factor test, the concrete should freely fall from one hopper to another hopper without any external effort. During experimental work, in some tests a little external effort is required to make the concrete fall from one hopper to another hopper. In the present work, in all concrete mixes, the compaction factor was in normal range for the concrete mixes containing 8 and 12 M of sodium hydroxide and sodium silicate to sodium hydroxide ratio of 2 and 3 for plastic up to 1%, but at 1.5% metallized polymer plastic, there is slight reduction in the compaction factor even though concrete was in workable condition to required degree of compaction (Figures 7.3 and 7.4).

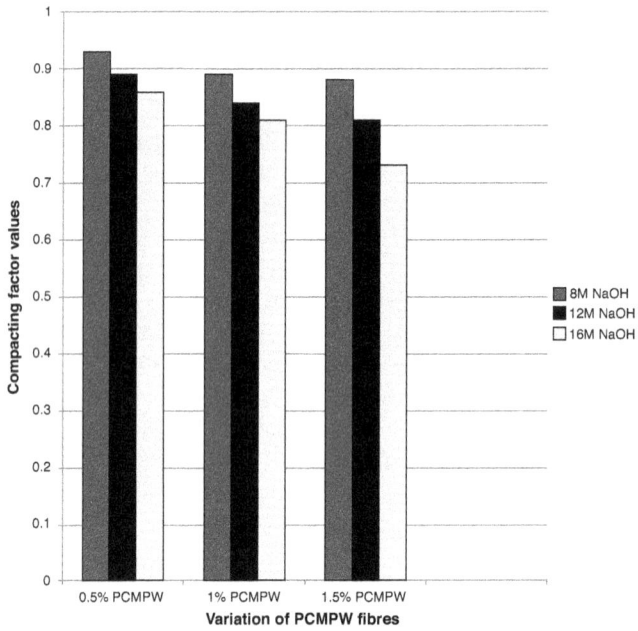

FIGURE 7.3   Effect of oxide ratio and plastic fibers on compacting factor.

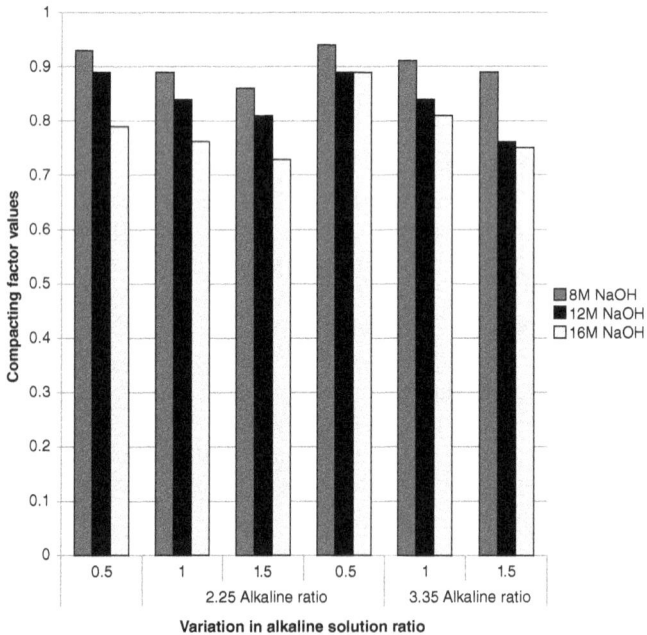

FIGURE 7.4   Effect of oxide ratio and plastic fibers on the compacting factor.

From the results, it may be observed that the compaction ability of the mixes of GPC consisting of varying fractions of the PCMPW is not significantly affected. The reason for this response is the compressible nature of the fibers. The overall volume of the fiber in the mixes even at 1.0% of the total mix volume ranges between 15 and 20 kg/m³ by weight of the plastic. However, the weight of the fiber is not significant, as the volume is on the higher side. This scenario impacts the workability to a great extent. Conversely, the compaction ability is also influenced by the over-concentration of the plastic in the mixture. Therefore, it is advisable to use optimum quantity of the plastic in GPC or in cement-based concrete. The detailed impact of chemical and plastic fibers in GPC on compaction factor is presented here.

Effect of NaOH: With increase in molarity of sodium hydroxide (NaOH) solution, the compaction factor of concrete decreases. For a constant sodium silicate to sodium hydroxide ratio (Na_2SiO_3/NaOH), compaction factor of concrete decreased with the increase in the concentration of sodium hydroxide. When the NaOH was increased from 8 to 12 M, the compaction factor decreased by 4.3%–5.31%, 3.44%–5.61%, and 5.81%–7.59% and when the NaOH was increased from 12 to 16 M, the compaction factor decreased by 0%–1.12%, 0%–3.57%, and 0%–9.87% for the polymer plastic of 0.5%, 1%, and 1.5%, respectively, in case of SiO_2/Na_2O ratio of 2.25. When the NaOH was increased from 8 to 12 M, the compaction factor was decreased by 3.37%–4.91%, 1.19%–5.61%, and 0%–11.62% and when the NaOH was increased from 12 to 16 M, the compaction factor decreased by 0%–3.48%, 3.57%–3.61%, and 1.31%–3.89% for the polymer plastic of 0.5%, 1%, and 1.5%, respectively, in case of SiO_2/Na_2O ratio of 3.35.

Effect of Na_2SiO_3/NaOH: As the ratio of sodium silicate to sodium hydroxide increased, the compaction factor decreased in all the batches. For increase in Na_2SiO_3/NaOH ratio from 1.0 to 2.0 the compaction factor decreased by 0%–1.07%, 1%–2.25%, and 0%–8.13% and for increase in Na_2SiO_3/NaOH ratio from 2.0 to 3.0 the compaction factor decreased by 10.22%–11.33%, 6.17%–9.52%, and 0%–9.87% for the polymer plastic of 0.5%, 1%, and 1.5%, respectively, in case of SiO_2/Na_2O ratio of 2.25. For increase in Na_2SiO_3/NaOH ratio from 1.0 to 2.0 the compaction factor decreased by 0.42%–1.11%, 2.19%–5.61%, and 3.37%–8.43% and for increase in Na_2SiO_3/NaOH ratio from 2.0 to 3.0 the compaction factor decreased by 3.37%–6.74%, 1.19%–5.61%, and 1.33%–10.46% for the polymer plastic of 0.5%, 1%, and 1.5%, respectively, in case of SiO_2/Na_2O ratio of 3.35.

Effect of metallized polymer plastic: For a constant molar concentration of sodium hydroxide and constant Na_2SiO_3/NaOH ratio the compaction factor of the concrete decreased with the increase of metallized polymer plastic. When metallized plastic waste increased from 0.5% to 1% and 0.5% to 1.5%, the reduction in the compaction factor was in the range of 3.61%–7.95% and 4.81%–17.04% for SiO_2/Na_2O ratio of 2.25 and 1.11%–8.98% and 5.31%–15.73% for SiO_2/Na_2O ratio of 3.35.

7.1.2.3 Density of the Fresh GPC Mixes

In addition to the slump and compacting factor values, the density of the mix should also be investigated. This can be an important study for any new composite material being developed using non-conventional material, especially when the mixture consists of the variation of the constituents and different combinations thereof. Therefore, the density aspects are also presented here for general discussion.

Effect of NaOH: With increase in molarity of sodium hydroxide (NaOH) solution, the density of concrete increases. For a constant sodium silicate to sodium hydroxide ratio (Na_2SiO_3/NaOH), the density of concrete increased with the increase in the concentration of sodium hydroxide. When the NaOH was increased from 8 to 12 M, the density also increased by 0.12%–0.66%, 0.16%–1.01%, and 0%–1.36% and when the NaOH was increased from 12 to 16 M, the density increased by 0.73%–1.1%, 0.41%–0.49%, and 0%–0.83% for the polymer plastic of 0.5%, 1%, and 1.5%, respectively, in case of SiO_2/Na_2O ratio of 2.25. When the NaOH was increased from 8 to 12 M, the density increased by 0.08%–2.96%, 0.33%–1.57%, and 0.08%–2.02% and when the NaOH was increased from 12 to 16 M, the density increased by 0.28%–0.69%, 0.33%–0.58%, and 1%–1.48% for the polymer plastic of 0.5%, 1%, and 1.5%, respectively, in case of SiO_2/Na_2O ratio of 3.35.

Effect of Na_2SiO_3/NaOH: As the ratio of sodium silicate to sodium hydroxide increased, the density also increased. For increase in Na_2SiO_3/NaOH ratio from 1.0 to 2.0 the density increased by 0.37%–0.78%, 0.79%–1.22%, and 0.41%–1.48% and for increase in Na_2SiO_3/NaOH ratio from 2.0 to 3.0 the density increased by 0.08%–0.57%, 0.24%–1.16%, and 0.33%–1.17% for the polymer plastic of 0.5%, 1% and 1.5%, respectively, in case of SiO_2/Na_2O ratio of 2.25. While for increase in Na_2SiO_3/NaOH ratio from 1.0 to 2.0 the density increased by 0.28%–3.05%, 0.62%–1.87%, and 0.5%–1.42% and for increase in Na_2SiO_3/NaOH ratio from 2.0 to 3.0 the density increased by 0.36%–0.77%, 0.45%–0.70%, and 0.36%–0.77% for the polymer plastic of 0.5%, 1%, and 1.5%, respectively, in case of SiO_2/Na_2O ratio of 3.35.

Effect of metallized polymer plastic: For a constant molar concentration of sodium hydroxide and constant Na_2SiO_3/NaOH ratio the density of the concrete decreased with the increase of metallized polymer plastic. When metallized plastic waste increased from 0.5% to 1% and 0.5% to 1.5%, the reduction in the density was in the range of 1.22%–2.71% and 1.87%–3.04% for SiO_2/Na_2O ratio of 2.25 and 0.42%–1.76% and 1.82%–3.76% for SiO_2/Na_2O ratio of 3.35.

From all three experimental data on workability response of GPC consisting of plastic fibers, it was observed that the addition of the fibers tends to reduce the flow characteristics to an extent, however, only with the addition of fibers beyond 1% of the volume of the mixtures.

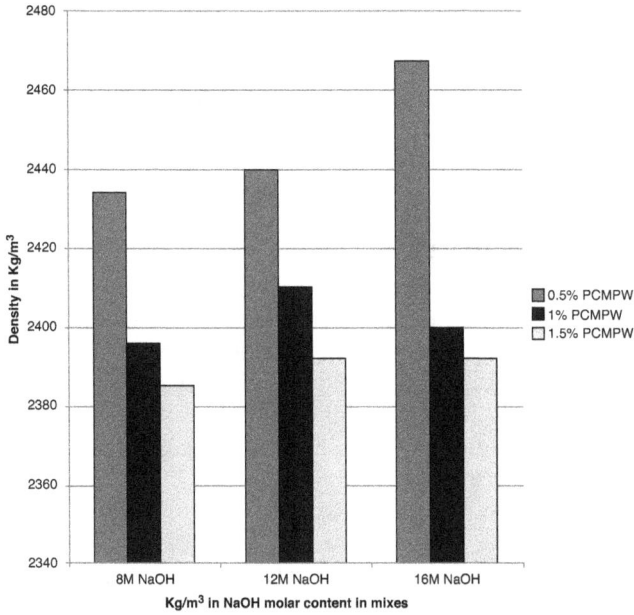

FIGURE 7.5   Effect of plastic fraction and oxide ratio on the compressive strength.

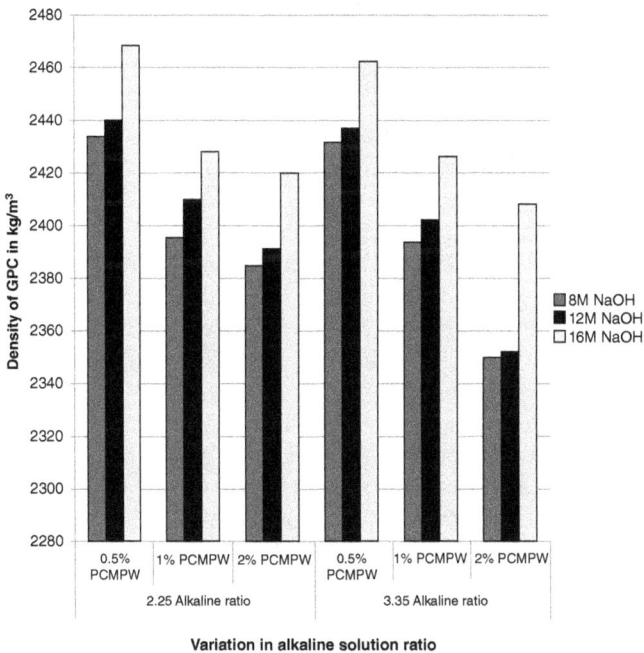

FIGURE 7.6   Effect of plastic fraction, NaOH molar, and oxide ratio on the compressive strength.

7.1.3 MECHANICAL PROPERTIES OF THE GPC CONSISTING OF PCMPW FIBERS

The mechanical properties of the GPC mixes include response of the material in hardened state to the compression, splitting actions, flexure deformations, and impact resistance. In this section, these properties are discussed in the form of experimental results. Kindly note that looking at the scope of the book, the results are shown as the sample data; however, their impacts on the properties have been discussed in detail.

7.1.3.1 Compressive Strength

Cubes of 150 mm × 150 mm × 150 mm size were cast and cured for 24 hours at 100°C in the laboratory oven. The necessary polymerization of the GPC mix takes place during the external heat curing. It is nowadays possible to develop the GPC without oven curing using the appropriate proportions of the silica fume and granulated blast furnace slag in powder form. The author will be publishing the literature on this segment soon with detailed mix design and method of preparing geopolymer without external heat curing of the specimens. However, the primary objective of this book is to provide the information about how to utilize industrial wastes in concrete, therefore the conventional method of producing geopolymer has been presented throughout the book.

Effect of NaOH: With increase in molarity of sodium hydroxide (NaOH) solution, the compressive strength of concrete increases. For a constant sodium silicate to sodium hydroxide ratio ($Na_2SiO_3/NaOH$), compressive strength of concrete increased with the increase in the concentration of sodium hydroxide. When the NaOH was increased from 8 to 12 M, the compressive strength increased by 3.65%–27.33%, 3.68%–12.43%, and 2.38%–18.35% and when the NaOH was increased from 12 to 16 M, the compressive strength increased by 3.86%–4.25%, 3.19%–10.17%, and 4.74%–5.81% for the polymer plastic of 0.5%, 1%, and 1.5%, respectively, in case of SiO_2/Na_2O ratio of 2.25. While, when the NaOH was increased from 8 to 12 M, the compressive strength increased by 14.56%–41.16%, 14.59%–38.47%, and 5.57%–19.15% and when the NaOH was increased from 12 to 16 M, the compressive strength increased by 12.39%–20.78%, 6.41%–7.28%, and 13.66%–16.93% for the polymer plastic of 0.5%, 1%, and 1.5%, respectively, in case of SiO_2/Na_2O ratio of 3.35.

Effect of $Na_2SiO_3/NaOH$: As the ratio of sodium silicate to sodium hydroxide increased, the compressive strength also increased. For increase in $Na_2SiO_3/NaOH$ ratio from 1.0 to 2.0 the compressive strength increased by 28.66%–35.90%, 19.33%–25.67%, and 4.57%–20.88% and for increase in $Na_2SiO_3/NaOH$ ratio from 2.0 to 3.0 the compressive strength increased by 7.50%–25.50%, 11.21%–22.25%, and 10.60%–22.58% for the polymer plastic of 0.5%, 1%, and 1.5%, respectively, in case of SiO_2/Na_2O ratio of 2.25. While, for increase in $Na_2SiO_3/NaOH$ ratio from 1.0 to 2.0 the compressive strength increased by 1.35%–24.35%, 0.68%–5.11%, and 2.08%–3.49%, and for increase in $Na_2SiO_3/NaOH$ ratio from 2.0 to 3.0 the compressive strength increased by 5.50%–13.37%, 1.32%–18.23%, and 2.15%–18.60% for the polymer plastic of 0.5%, 1%, and 1.5%, respectively, in case of SiO_2/Na_2O ratio of 3.35.

Effect of metallized polymer plastic: For a constant molar concentration of sodium hydroxide and constant $Na_2SiO_3/NaOH$ ratio the compressive strength of the concrete decreased with the increase in metallized polymer plastic. When metallized plastic waste increased from 0.5% to 1% and 0.5% to 1.5%, the reduction in the compressive strength was in the range of 3.26%–18.06% and 8.04%–28.53% for SiO_2/Na_2O ratio of 2.25 and 2.97%–19.34% and 4.89%–24.21% for SiO_2/Na_2O ratio of 3.35.

The results suggested that the molar concentration of NaOH plays a key role in strength gaining mechanism of the GPC. Though the plastic fibers do not remain chemically active, they impose limitations on lack of coherence of the constituents when more than 1% of the volume of the mix is added. One of the optimum mixtures using GPC and PCMPW may be prepared with 16 M molar concentration and 1% plastic fiber by volume to take maximum advantage of waste utilization in concrete preparation. Nevertheless, the wastes may be used with maximum proportion of plastics and chemical composition for non-structural purpose also.

7.1.3.2 Splitting Tensile Strength

The response by the GPC mixes consisting of plastic waste was found to be excellent in terms of resistance to the splitting action applied on the specimens. This is with respect to the conventional concrete of the same target strength and consisting of plastic fibers as discussed in Chapter 5. The average values of the splitting tensile strength of the mixes are regarded with the compressive strength of the concrete. In the case of GPC prepared using only wastes without inclusion of plastics as discussed in Chapter 6, we have observed that the average value is reported to be higher than ordinary concrete. The same has been discussed here; however, the impact of addition of the plastic fibers has been focused. An additional discussion is the influence of the aspect ratio as the size of the PCMPW fibers on the results of the splitting resistance test of the specimens. The test results are shown in Figure 7.7.

From the above figure, several information regarding the performance of the GPC with plastic may be obtained. The results indicate the combined information about the normal GPC concrete prepared with 8 M NaOH molarity, and the oxide ratio of 2 and the liquid to fly ash ratio of 0.45, resembling the M20 grade of normal

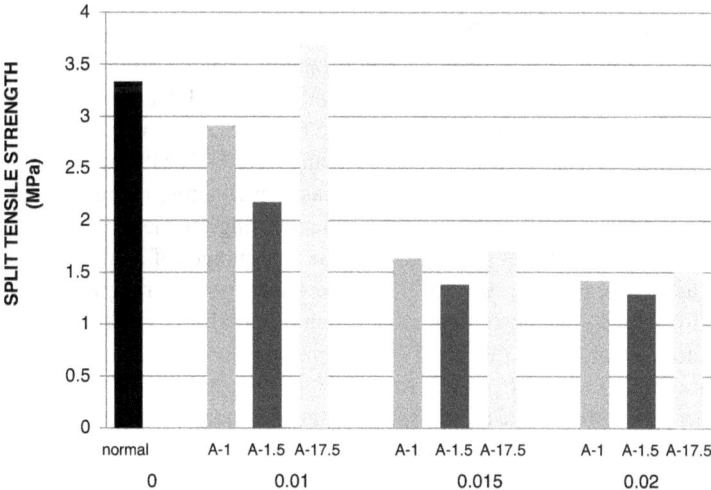

FIGURE 7.7 Variation of the splitting tensile strength due to plastic fibers and sizes.

cement-based concrete, addition, and variation of the volumetric fraction inclusion of PCMPW fibers, and the aspect ratio of the fibers.

Effect of NaOH: With increase in molarity of sodium hydroxide (NaOH) solution, the splitting tensile strength of concrete increases. For a constant sodium silicate to sodium hydroxide ratio (Na_2SiO_3/NaOH), splitting tensile strength of concrete increased with the increase in the concentration of sodium hydroxide. When the NaOH was increased from 8 to 12 M, the splitting tensile strength increased by 10.33%–46.28%, 2.64%–26.59%, and 7.92%–32.24% and when the NaOH was increased from 12 to 16 M, the splitting tensile strength increased by 12.80%–24.34%, 21.89%–25.72%, and 27.91%–30.30% for the polymer plastic of 0.5%, 1%, and 1.5%, respectively, in case of SiO_2/Na_2O ratio of 2.25. While, when the NaOH was increased from 8 to 12 M, the splitting tensile strength increased by 29.31%–38.43%, 23.96%–50.26%, and 12.44%–50.35% and when the NaOH was increased from 12 to 16 M, the splitting tensile strength increased by 9.39%–16.71%, 9.05%–28.33%, and 14.15%–18.70% for the polymer plastic of 0.5%, 1%, and 1.5%, respectively, in case of SiO_2/Na_2O ratio of 3.35.

Effect of Na_2SiO_3/NaOH: As the ratio of sodium silicate to sodium hydroxide increased, the splitting tensile strength also increased. For increase in Na_2SiO_3/NaOH ratio from 1.0 to 2.0 the splitting tensile strength increased by 9.50%–27.71%, 8.81%–33.47%, and 5.94%–29.81% and for increase in Na_2SiO_3/NaOH ratio from 2.0 to 3.0 the splitting tensile strength increased by 6.79%–21.40%, 5.37%–8.68%, and 4.94%–16.35% for the polymer plastic of 0.5%, 1%, and 1.5%, respectively, in case of SiO_2/Na_2O ratio of 2.25. While for increase in Na_2SiO_3/NaOH ratio from 1.0 to 2.0, the splitting tensile strength increased by 26.17%–33.60%, 12.35%–31.05%, and 6.60%–20.56% and for increase in Na_2SiO_3/NaOH ratio from 2.0 to 3.0 the splitting tensile strength increased by 5.80%–14.12%, 4.52%–23%, and 15.92%–37.05% for the polymer plastic of 0.5%, 1%, and 1.5%, respectively, in case of SiO_2/Na_2O ratio of 3.35.

Effect of metallized plastic fibers: For a constant molar concentration of sodium hydroxide and constant Na_2SiO_3/NaOH ratio the splitting tensile strength of the concrete decreased with the increase of metallized polymer plastic. When metallized plastic waste increased from 0.5% to 1% and 0.5% to 1.5%, the reduction in the splitting tensile strength was in the range of 5.65%–18.35% and 12.01%–28.26% for SiO_2/Na_2O ratio of 2.25 and 5.09%–20.74% and 8.62%–31.51% for SiO_2/Na_2O ratio of 3.35. The addition of metallized plastic waste in concrete decreases the splitting tensile strength in aspect ratio 1 and 1.5. The splitting tensile strength increases in aspect ratio 17.5 up to 1% of plastic waste. The percentages of addition of metallized plastic waste increase from 0% to 2%, while compressive strength decreases from 012.61% to 47.64%. The percentage of addition of metallized plastic waste increases from 0% to 1% (for aspect ratio 17.5) as compared to normal concrete. The splitting tensile strength increases from 0% to 10.81% as compared to normal concrete test result. For 1% of plastic waste, the splitting tensile strength decreases from 12.61% to 34.83%. For 1.5% of plastic waste, the splitting tensile strength decreases from 38.14% to 42.64%. For 2% of plastic waste, the splitting tensile strength decreases from 43.24% to 47.45%. The aspect ratios considered were 1, 1.5, and 17.5. By changing the aspect ratio, the splitting tensile response also changes. The minimum splitting tensile strength decreased in aspect ratio with 1% of plastic waste. The maximum reduction was in aspect ratio 1.5% and 2% of plastic waste. For aspect ratio of 1%, the splitting tensile strength decreases from 12.61% to 43.24%, and for aspect ratio 1.5, the reduction was from 34.83% to 47.45% and for 17.5, the splitting tensile strength increases 10.81 (1% of plastic).

7.1.3.3 Flexure Strength

The strength under flexure actions is of prime importance because the structural members made with concrete show brittle behavior generally in case of no reinforcement provided in the matrix. The fibers added are expected to add ductility or elasticity to the material. The two-point or one-point loading conditions are generally employed to carry out the flexure strength calculation on the members. Here the test results have been discussed for specimens of 100 mm × 100 mm × 70 mm size beam members subjected to compression flexure actions. The effect of GPC constituents is discussed in detail.

Generally, the flexure strength modulus and the compressive strength or flexural modulus and splitting tensile strength of the concrete are compared for establishing inter-relationships. The material is evaluated for resistance against the cracking resistance and crack propagation under transverse loading to the principal axis of the members.

7.1.3.4 Impact Strength

As discussed in the earlier chapters, the impact resistance is an important response to be observed under sudden loading conditions on the concrete. The major limitations of conventional unreinforced concrete are non-homogeneity or scattered physical properties of the matrix, limited cracking resistance and non-uniform stress distribution within the hardened mass against axial loading. The impact test was employed to compare the behavior of the material against surface damage and energy absorption capacity. Also, the inclusion of the fibers is required to be evaluated. Following are the test results and discussion on the experimental work carried out on disk specimens under the free falling 4.5 kg weight from the 450 mm distance repeatedly applied on the same location.

All three varying molar values of NaOH and the corresponding fraction of the plastic fibers have been analyzed. The results are shown in the form of the total number of blows required to cause the final failure of the members. The sample results are shown in Figure 7.8.

Effect of NaOH: As the molarity of sodium hydroxide (NaOH) solution increases, the impact resistance of concrete increases. For a constant sodium silicate to sodium hydroxide ratio (Na_2SiO_3/NaOH), impact resistance of concrete increased with the increase in the concentration of sodium hydroxide. When the NaOH was increased from 8 to 12 M, the impact resistance increased by 17.90%–50.97%, 5.66%–32.29%, and 16.91%–45.91% and when the NaOH was increased from 12 to 16 M, the impact resistance increased by 9.27%–13.47%, 23.06%–92.29%, and 10.77%–26.04% for the polymer plastic of 0.5%, 1% and 1.5%, respectively, in case of SiO_2/Na_2O ratio of 2.25. While when the NaOH was increased from 8 to 12 M, the impact resistance increased by 15.24%–20.65%, 15.70%–46.86%, and 26.20%–55.17% and when the

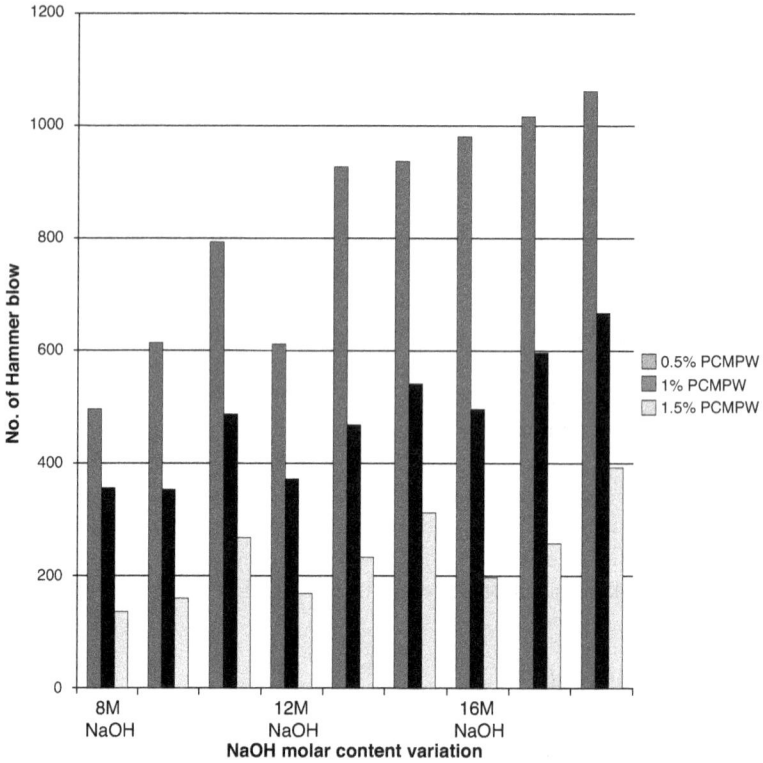

FIGURE 7.8 Effect of oxide ratio and plastic fiber fraction on impact resistance.

NaOH was increased from 12 to 16 M, the impact resistance increased by 9.82%–12.25%, 14.25%–14.82%, and 8.19%–33.07% for the polymer plastic of 0.5%, 1%, and 1.5%, respectively, in case of SiO_2/Na_2O ratio of 3.35.

Effect of $Na_2SiO_3/NaOH$: As the ratio of sodium silicate to sodium hydroxide increased, the impact resistance also increased. For increase in $Na_2SiO_3/NaOH$ ratio from 1.0 to 2.0 the impact resistance increased by 23.79%–51.71%, 0%–25.20%, and 17.77%–38.09% and for increase in $Na_2SiO_3/NaOH$ ratio from 2.0 to 3.0 the impact resistance increased by 0.86%–29.15%, 16.06%–37.96%, and 34.05%–67.29% for the polymer plastic of 0.5%, 1%, and 1.5%, respectively, in case of SiO_2/Na_2O ratio of 2.25. While for increase in $Na_2SiO_3/NaOH$ ratio from 1.0 to 2.0 the impact resistance increased by 8.45%–12.07%, 25.25%–73.71%, and 86.52%–157.5% and for increase in $Na_2SiO_3/NaOH$ ratio from 2.0 to 3.0 the impact resistance increased by 13.78%–19.12%, 10.5%–40.95%, and 12.5%–40.77% for the polymer plastic of 0.5%, 1%, and 1.5%, respectively, in case of SiO_2/Na_2O ratio of 3.35.

Effect of metallized polymer plastic: For a constant molar concentration of sodium hydroxide and constant $Na_2SiO_3/NaOH$ ratio the impact resistance of the concrete decreased with the increase of metallized polymer plastic. When metallized plastic waste increased from 0.5% to 1% and 0.5% to 1.5%, the reduction in the impact resistance was in the range of 11.35%–49.62% and 63.05%–74.97% for $SiO_2/$

Na_2O ratio of 2.25 and 34.04%–67.50% and 74.51%–91.66% for SiO_2/Na_2O ratio of 3.35, respectively.

Overall response by the disk specimens prepared with the GPC and plastic fibers was superior compared to the conventional concrete and the plain GPC mixes. The combination of GPC with fibers increased the strength of the member significantly and showed better stress distribution and resistance against the failure of the member. The fibers were able to resist the crack propagation in the mass.

7.1.3.5 Rebound Hammer Response

As a confirmation of the compressive strength and the coherence of the material in the hardened state, one of the most commonly used tests on concrete is the rebound hammer test. It is a non-destructive test wherein the members are tested with the automatic rebound hammer with digital gauge showing the number of rebounds obtained by the material, and the corresponding strength values from the standard relationship is measured. Though the rebound hammer test possesses an approximation and cannot be used as the only method to measure the strength of the material, it provides nearly acceptable results provided that the hammer is placed and applied appropriately. Figure 7.9 is an image of a rebound hammer.

It consists of a spring control hammer that slides on a plunger within a tabular housing. Upon pressing the hammer against the surface of the concrete, the mass rebounds from the plunger. It retracts against the force of the spring. The rider is carried along the guidance scale by the spring control mass as it rebounds after the hammer strikes the concrete. By pushing a button, the rider can be held in position to allow the reading to be taken. The distance traveled by the mass is the rebound number. It is indicated by the rider moving along a graduated scale.

For use on concrete made with aggregates from a particular supplier, each hammer needs to be calibrated due to its wide range of performance. The test can be conducted horizontally, vertically upward, or onward or at any intermediate angle. At each angle the rebound number will be different for the same concrete and will

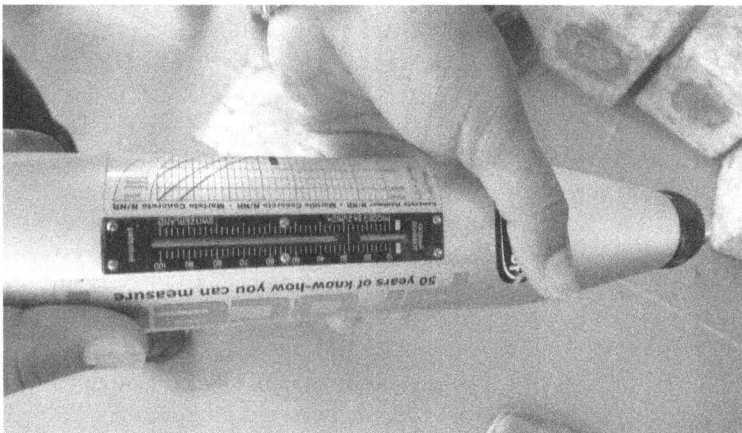

FIGURE 7.9   Rebound hammer apparatus.

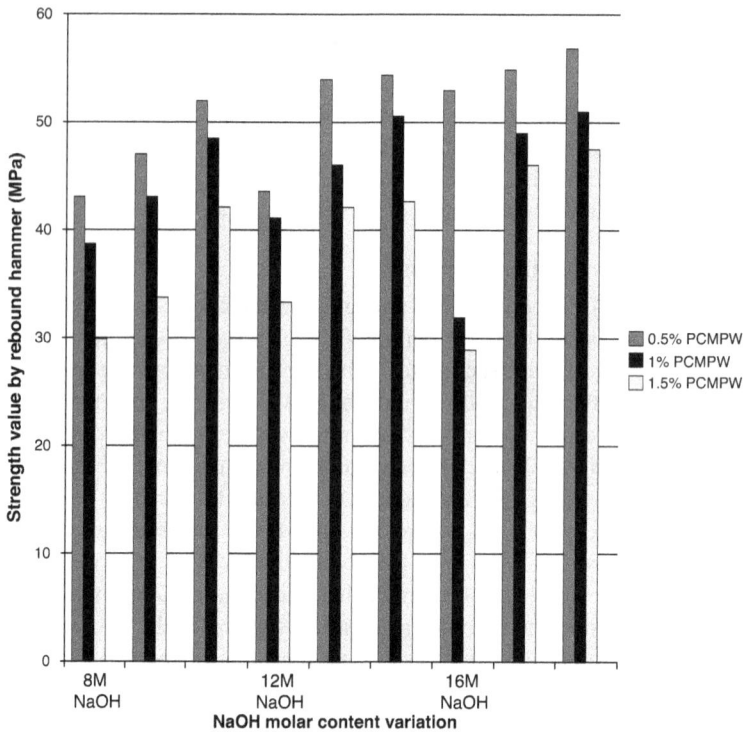

FIGURE 7.10   Effect of oxide ratio and plastic fibers on rebound hammer response.

require separate calibration or correction chart. A calibration chart is available with the instrument (Figure 7.10).

Effect of NaOH: As the molarity of sodium hydroxide (NaOH) solution increases, the surface strength of concrete measured by Schmidt's rebound hammer increases. For a constant sodium silicate to sodium hydroxide ratio ($Na_2SiO_3/NaOH$), surface strength of concrete increased with the increase in the concentration of sodium hydroxide. When the NaOH was increased from 8 to 12 M, the average rebound number was increased by 0.43%–9.75%, 1.57%–5.17%, and 1.31%–10.15% and when the NaOH was increased from 12 to 16 M, the average rebound number was increased by 3.33%–5.54%, 0.76%–4.92%, and 7.02%–8.2% for the polymer plastic of 0.5%, 1%, and 1.5%, respectively, in case of SiO_2/Na_2O ratio of 2.25. While when the NaOH was increased from 8 to 12 M, the average rebound number increased by 1.34%–9.24%, 0.82%–7.53%, and 1.55%–9.54% and when the NaOH was increased from 12 to 16 M, the average rebound number increased by 0.76%–1.14%, 1.29%–2.86%, and 2.75%–4.95% for the polymer plastic of 0.5%, 1%, and 1.5%, respectively, in case of SiO_2/Na_2O ratio of 3.35.

Effect of $Na_2SiO_3/NaOH$: As the ratio of sodium silicate to sodium hydroxide increased, the surface strength of concrete measured by Schmidt's rebound hammer also increased. For increase in $Na_2SiO_3/NaOH$ ratio from 1.0 to 2.0 the average rebound number increased by 6.03%–15.87%, 8.43%–8.92%, and 4.02%–11.21% and for increase in $Na_2SiO_3/NaOH$ ratio from 2.0 to 3.0 the average rebound number increased by 0.37%–6.91%, 1.54%–9.48%, and 1.31%–10.14% for the polymer plastic of 0.5%, 1%, and 1.5%, respectively, in case of SiO_2/Na_2O ratio of 2.25. While, for increase in $Na_2SiO_3/NaOH$ ratio from 1.0 to 2.0 the average rebound number increased by 6.72%–15.04%, 8.41%–13.06%, and 3.10%–11.22% and for increase in $Na_2SiO_3/NaOH$ ratio from 2.0 to 3.0 the average rebound number increased by 0.38%–7.56%, 5.17%–7.55%, and 1.83%–5.02% for the polymer plastic of 0.5%, 1%, and 1.5%, respectively, in case of SiO_2/Na_2O ratio of 3.35.

Effect of metallized polymer plastic: For a constant molar concentration of sodium hydroxide, and constant $Na_2SiO_3/NaOH$ ratio, the surface strength of concrete measured by Schmidt's rebound hammer decreased with the increase of metallized polymer plastic. When metallized plastic waste increased from 0.5% to 1% and 0.5% to 1.5%, the reduction in the Schmidt's rebound number was in the range of 3.41%–9.10% and 12.01%–15.85% for SiO_2/Na_2O ratio of 2.25 and 4.92%–10.77% and 11.74%–18.37% for SiO_2/Na_2O ratio of 3.35.

7.1.3.6 Pull-Off Strength

A unique test for evaluation of the intermolecular bonding of the concrete composite is the pull-off strength test. This is an important parameter for the adhesion capacity of the intermolecular bonding of the mass in the matrix, especially in case of the applied external adhesive on the concrete surface. However, the test may be utilized to indicate the macrostructural response of the material possessing the fibers. The tests were performed on the GPC concrete cubes and with the help of the pull-off apparatus. The apparatus and method are briefly described here with illustrations.

Pull-off test is performed on concrete surface to measure the surface strength of concrete. In pull-off test, a direct tensile force is applied perpendicular to the surface of concrete. Surface strength is also known as bond strength. Pull-off test allows the measurement of the resistance of pull-off substrata of a portion of concrete surface. The tensile force is applied in a gradual manner without jump by a hydraulic force multiplier. The tensile force is applied to a metallic disk glued to the concrete surface. The result of the test is the tensile resistance to the pull-off of the concrete core under examination calculated as the ratio between failure load and the surface area of the metal disk used in the test. Pull-off test was performed on concrete surface after 7 days of casting. Size of the concrete specimen used in pull-off test is 150 mm × 150 mm × 150 mm cube. A core of 50 mm diameter was drilled in concrete specimen up to 50 mm depth as shown in Figure 5.12. The metal disk was glued on the concrete surface. After 24 hours, the bolt with bull head was inserted in the metal disk glued on the concrete surface. The side lever of pull-off testing machine was rotated to align the index to the maximum value of 60 mm. The pull-off testing apparatus was started and the digital indicator of pull-off test apparatus was made zero. The pull-off test apparatus was placed on the concrete surface as much parallel as possible with respect to the concrete surface. The head of bolt was fitted into the bottom seating arrangement of pull-off apparatus. The large central hand wheel was tightened until a slight tension pressure is applied. The load applicator lever of pull-off test apparatus was slowly rotated. The metal disk was pulled off from the concrete surface with the concrete core at failure load. The failure pull-off load of specimen as well as failure pull-off strength in MPa were recorded. Both readings are displayed separately on the display screen. During the pull-off test on concrete specimens it was observed that in concrete specimens which were cast from geopolymer concrete using metallized polymer plastic, the 30–50 mm core of concrete comes out with metal disk from concrete specimen at failure load (Figure 7.11).

The test results are shown in Figure 7.12. The results were analyzed for the influence of the GPC and fibers' internal adhesion and bonding as well as resistance against the direct pull of the material.

Effect of NaOH: As the molarity of sodium hydroxide (NaOH) solution increases, the pull-off strength of concrete increases. For a constant sodium silicate to sodium hydroxide ratio (Na_2SiO_3/NaOH), pull-off strength of concrete increased with the increase in the concentration of sodium hydroxide. When the NaOH was increased from 8 to 12 M, the pull-off strength was increased by 44%–55.58%, 33.33%–40.48%, and 11.56%–30.57% and when the NaOH was increased from 12 to 16 M, the pull-off strength increased by 2.23%–2.64%, 16.99%–20.37%, and 17.17%–29.85% for the polymer plastic of 0.5%, 1%, and 1.5%, respectively, in case of SiO_2/Na_2O ratio of 2.25. While, when the NaOH was increased from 8 to 12 M, the pull-off strength increased by 34.16%–50.84%, 18.20%–32.69%, and 14.82%–55.17% and when the NaOH was increased from 12 to 16 M, the pull-off strength increased by 14.18%–27.01%, 14.09%–25.44%, and 4.16%–13.11% for the polymer plastic of 0.5%, 1%, and 1.5%, respectively, in case of SiO_2/Na_2O ratio of 3.35.

Effect of Na_2SiO_3/NaOH: As the ratio of sodium silicate to sodium hydroxide increased, the pull-off strength also increased. For increase in Na_2SiO_3/NaOH ratio from 1.0 to 2.0 the pull-off strength increased by 17.81%–27.29%, 4.09%–4.27%, and 7.80%–11.91% and for increase in Na_2SiO_3/NaOH ratio from 2.0 to 3.0 the pull-off strength increased by 6.88%–14.07%, 6.90%–12.45%, and 4.78%–18.13% for the polymer plastic of 0.5%, 1%, and 1.5%, respectively, in case of SiO_2/Na_2O ratio of

FIGURE 7.11 Pull-off test apparatus and tested samples.

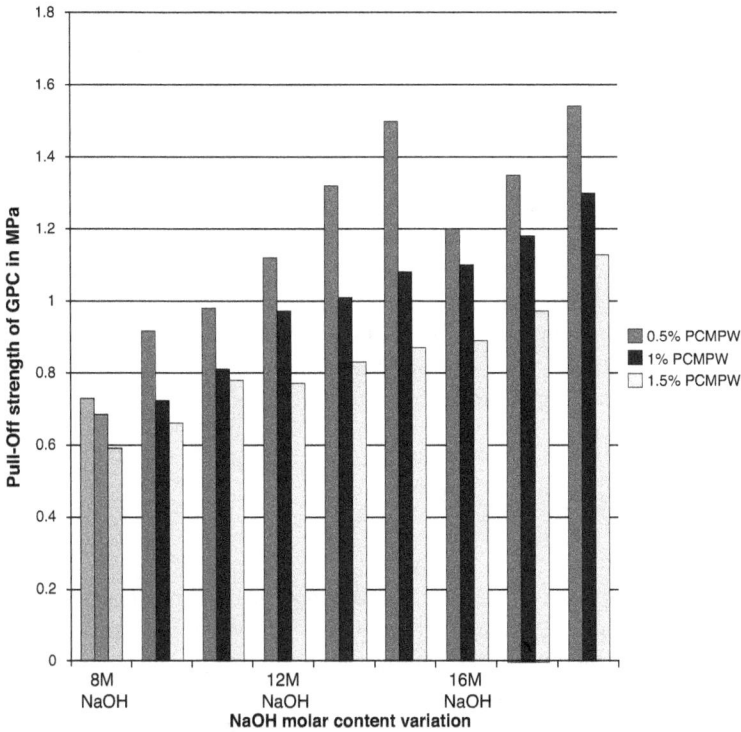

FIGURE 7.12   Effect of oxide and fraction of plastic fibers on pull-off strength response.

2.25. While for increase in $Na_2SiO_3/NaOH$ ratio from 1.0 to 2.0 the pull-off strength increased by 10.75%–22.89%, 11.01%–24.61%, and 3.25%–24.13% and for increase in $Na_2SiO_3/NaOH$ ratio from 2.0 to 3.0 the pull-off increased by 8.36%–20.54%, 14.58%–36.21%, and 8.32%–21.75% for the polymer plastic of 0.5%, 1%, and 1.5%, respectively, in case of SiO_2/Na_2O ratio of 3.35.

Effect of metallized polymer plastic: For a constant molar concentration of sodium hydroxide, and constant $Na_2SiO_3/NaOH$ ratio the pull-off strength of the concrete decreased with the increase of metallized polymer plastic. When metallized plastic waste increased from 0.5% to 1% and 0.5% to 1.5%, the reduction in the pull-off strength was in the range of 4.10%–28.01% and 18.10%–42% for SiO_2/Na_2O ratio of 2.25 and 11.66%–26.51% and 21.72%–40.34% for SiO_2/Na_2O ratio of 3.35, respectively.

From the test results of strength properties of concrete prepared with the waste, namely GPC, and consisting of plastic fibers, it may be observed that the addition of the fibers increases the crack resistance and crack propagation of the concrete mass. Here the concrete is also prepared from the GPC as the material had improved intermolecular bonding of the constituents. The general observation of the results suggested that all the strength properties of GPC with plastic waste fibers are on the higher side provided that the maximum addition of the fibers is maintained up to 1%

by volume of the mixture. The next section deals with the durability aspects of the GPC consisting of the fibers rendered from the metallized plastic wastes.

7.2 DURABILITY PROPERTIES

Durability properties are referred as the overall resistance of the material to the environmental conditions and external forces other than the structural loads on the concrete. The primary loading conditions such as acid and sulfate attacks, corrosion of concrete, and chloride penetration scenario of concrete are referred as the durability resistance of the concrete. Like the earlier Chapters 5 and 6, the results of the test conducted on various specimens prepared with the GPC consisting of the PCMPW fibers have been presented and discussed for obtaining the overall scenario on how the GPC and plastic waste fibers may resist the impacts of external and internal degradation of the concrete.

7.2.1 Acid and Sulfate Attack

7.2.1.1 Acid Test

The cubes were cast at size of 150 mm × 150 mm × 150 mm and kept at a temperature of 100°C at the oven for 24 hours. After 24 hours the cubes were removed from the mold and put in normal room temperature. After 7 days the cubes were immersed in a 5% concentric sulfuric acid (H_2SO_4) and 5% hydrochloric acid (HCl). After 28 days of curing, the weight difference and the compressive strength of cubes were measured.

7.2.2 Sulfate Test

The test procedure for sulfate resistance was developed by modifying the related Standards for normal Portland cement and concrete (Standards ASTM, 1993, 1995, 1997; Standards-Australia, 1996b). A total of nine cube specimen are kept in normal room temperature for 7 days after casting. After 7 days, of the nine cubes, three cubes were kept in a dry place and their weight was taken and then the cubes were immersed in sulfate curing tank for up to 28 days. After 28 days the cubes were taken out and kept in a dry place and the weight difference and compressive strength of specimens were measured.

The cubes were cast at a size of 150 mm × 150 mm × 150 mm and kept at a temperature of 100°C in the oven for 24 hours. After 24 hours the cubes were removed from the mold and kept in normal room temperature. After 7 days curing the cubes were immersed in a 10% sodium sulfate solution. After 28 days of curing, the weight difference and compressive strength of cubes were measured (Figures 7.13–7.15).

7.2.2.1 Effect of NaOH

When the molar content of sodium hydroxide (NaOH) solution is increased results show that the weight difference in acid attack and sulfate attack is reduced. In acid curing, increasing the molarity of NaOH led to a reduction in the weight difference at 24.89%–56.95% in SiO_2/Na_2O of 2.25 and 23.08%–45.04% in SiO_2/Na_2O of 3.35. In

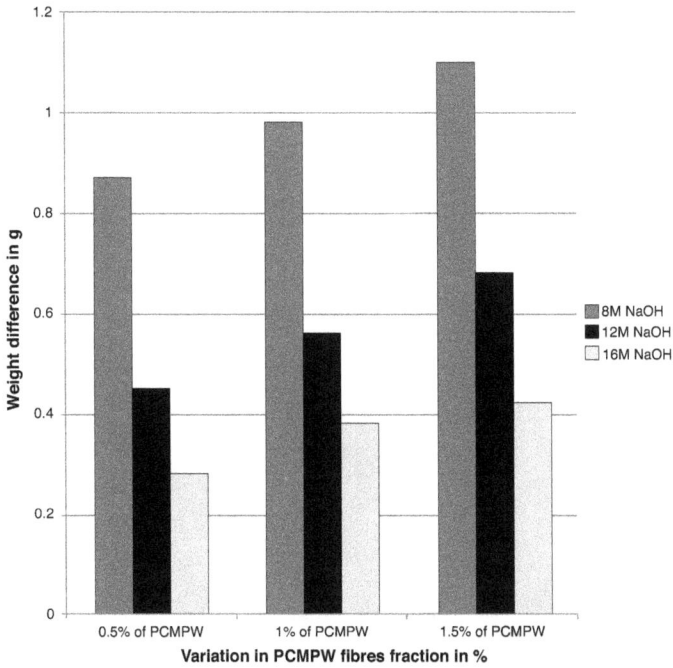

FIGURE 7.13   Effect of silicate to hydroxide ratio 1 on acid resistance of GPC.

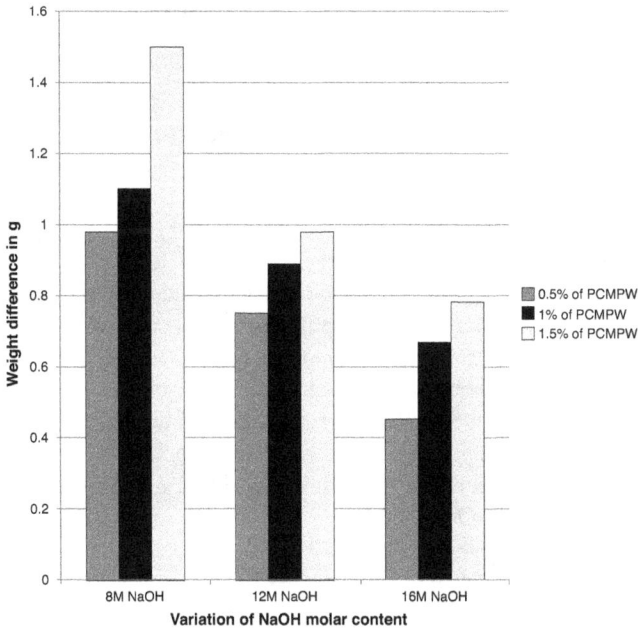

FIGURE 7.14   Effect of silicate to hydroxide ratio 2 on acid resistance of GPC.

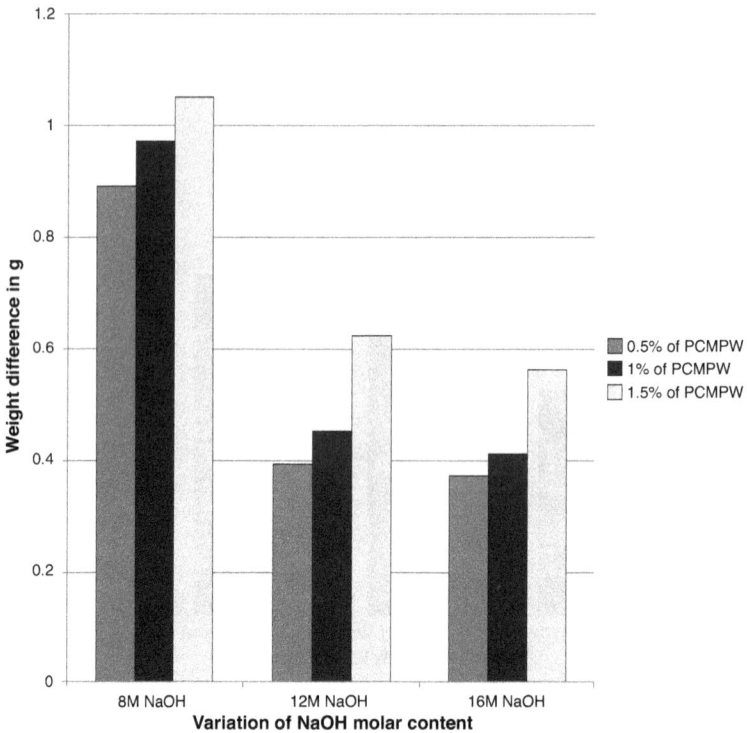

FIGURE 7.15   Effect of silicate to hydroxide ratio 3 on acid resistance of GPC.

sulfate curing, increasing the molarity of NaOH led to a reduction in the weight difference at 19.70%–59.93% in SiO_2/Na_2O of 2.25 and 25.92%–41.70% in SiO_2/Na_2O of 3.35. The test results indicate that weight was decreased at 5.98%–32.06% in acid curing compared to sulfate curing.

7.2.2.2   Effect of $Na_2SiO_3/NaOH$

When the ratio of sodium silicate to sodium hydroxide was increased, it led to decrease in the weight difference of geopolymer concrete. In acid curing, when the ratio of sodium silicate to sodium hydroxide increased from 1 to 3 then weight difference of geopolymer concrete decreased from 1.02% to 12.53% in SiO_2/Na_2O of 2.25 and 1.36%–28.54% in SiO_2/Na_2O of 3.35. In sulfate curing, when the ratio of sodium silicate to sodium hydroxide increased from 1 to 3 then the weight difference of geopolymer concrete decreased from 1.38% to 14.44% in SiO_2/Na_2O of 2.25 and 1.58%–15.75% in SiO_2/Na_2O of 3.35.

7.2.2.3   Effect of Metallized Polymer Plastic

For a constant molar content of sodium hydroxide and constant $Na_2SiO_3/NaOH$ ratio, the weight difference of the concrete increased with the increase of metallized polymer plastic. In acid curing, the weight difference of the concrete increased from 2.33% to 36.11% in SiO_2/Na_2O ratio of 2.25 and 1.61% to 32.92% in SiO_2/Na_2O ratio of 3.35 when percentage of plastic increased from 0.5% to 1.5%. In sulfate curing, the weight difference of the concrete increased from 2.17% to 39.73% in SiO_2/Na_2O ratio of 2.25 and 2.93% to 28.80% in SiO_2/Na_2O ratio of 3.35 when plastic increased from 0.5% to 1.5%.

7.2.2.4   Effect of Type of Curing on Normal Concrete

M20 grade of concrete weight difference increased 11.13% in sulfate curing compared to acid curing. M25 grade of concrete weight difference increased 25.55% in sulfate curing compared to acid curing.

The test showed excellent response by the GPC containing plastic waste fibers against weight reduction due to acid attack and volume change due to sulfate attack. The denser and stronger intermolecular structure of the GPC along with the plastic fibers was capable of resisting the degradation of the concrete due to the ingress of the acid and sulfate into the hardened concrete mass.

7.2.3   OXYGEN PERMEABILITY OF GPC WITH PCMPW FIBERS

Oxygen permeability test was used to measure the oxygen permeability of concrete. The durability of structures made of reinforced concrete is conditioned by the mechanisms of movement of aggressive fluids within the micro-structure of the concrete. In this respect the concrete acts as a protection barrier for the steel reinforcing bars against potentially damaging O_2, CO_2, and H_2O. Oxygen is normally selected for permeability testing given its chemical inertia within the porous structure of concrete. This method cannot be used on samples saturated in water as well as samples with fissures that are samples with evident defects such as honeycombing, cracks, etc. The test result is the mean specific coefficient of oxygen permeability. The dimensions of concrete cylinder disks were 150 mm dia.×50 mm height. During the oxygen permeability test on concrete disk, it was observed that the amount of oxygen gas required for testing concrete disk is not fixed. The number of concrete disks tested in one small-sized oxygen cylinder is not fixed. The number of concrete disks tested in one small-sized oxygen cylinder depends on duration of testing, proper connection of rubber tube between oxygen cylinder valve and oxygen permeability test apparatus, oxygen gas pressure of cylinder, control gas pressure from control valve at oxygen cylinder, increase and control of oxygen gas pressure with increase as well as control valve at oxygen permeability test apparatus, and skill as well as experience of oxygen permeability apparatus operator. The test provides the result in the form of distance traveled by the soap bubble in the measuring tubes attached with the apparatus during the steady flow of the oxygen. Also, the capacity of concrete to allow the passage of oxygen under the controlled condition can be determined in terms of the pressure gauge as the quantity of oxygen passing through the concrete may be expressed in terms of rate of passing of the volume of the gas per unit time in seconds. The results are discussed in the sections below (Figures 7.16–7.18).

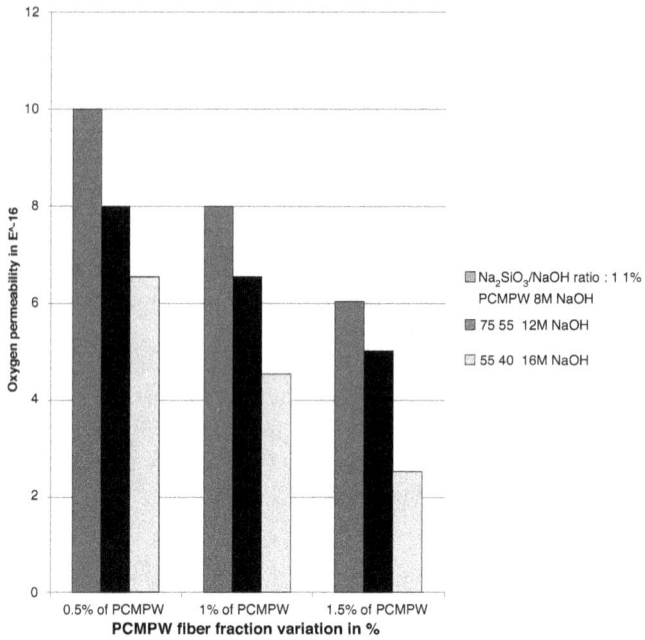

FIGURE 7.16   Effect of silicate to hydroxide ratio 1 on permeability of GPC.

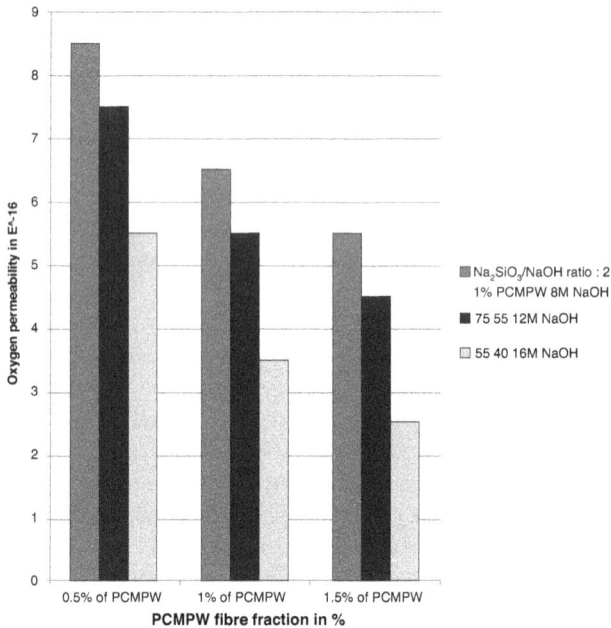

FIGURE 7.17   Effect of silicate to hydroxide ratio 2 on permeability of GPC.

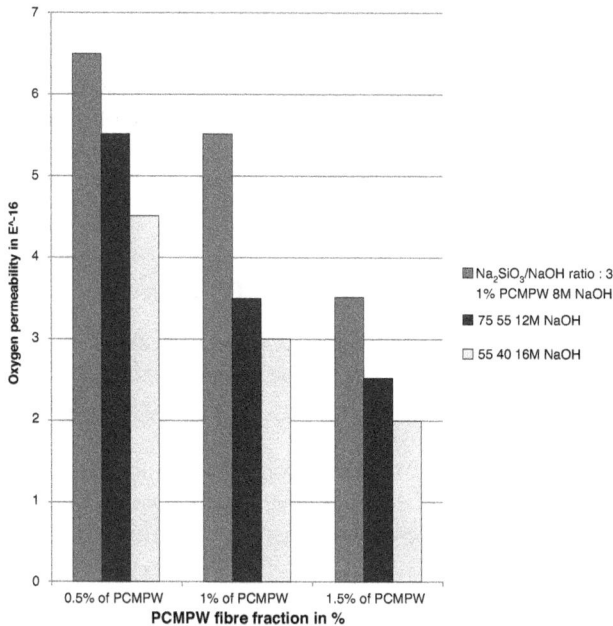

FIGURE 7.18 Effect of silicate to hydroxide ratio 3 on permeability of GPC.

7.2.3.1 Effect of NaOH

When the ratio of sodium silicate to sodium hydroxide was constant and the molar content of sodium hydroxide (NaOH) solution increased, results show that the oxygen permeability was reduced, as the molar content of NaOH was increased and then the oxygen permeability decreased from 17.91% to 52.61% in SiO_2/Na_2O of 2.25 and 12.81% to 26.51% in SiO_2/Na_2O of 3.35.

7.2.3.2 Effect of Na₂SiO₃/NaOH

When the molar content of sodium hydroxide (NaOH) solution was constant and the ratio of sodium silicate to sodium hydroxide was increased then the oxygen permeability of geopolymer concrete decreased. When the ratio of sodium silicate to sodium hydroxide increased from 1 to 3 then oxygen permeability of geopolymer concrete decreased from 0.06% to 19.44% in SiO_2/Na_2O of 2.25 and 0.71% to 9.42% in SiO_2/Na_2O of 3.35.

7.2.3.3 Effect of Metallized Polymer Plastic

For a constant molar content of sodium hydroxide and constant ratio of sodium silicate to sodium hydroxide, the oxygen permeability of geopolymer concrete decreased with the increase in metallized polymer plastic. Oxygen permeability decreased from 2.20% to 34.75% in SiO_2/Na_2O ratio of 2.25 and 0.96% to 13.84% in SiO_2/Na_2O ratio of 3.35 when percentage of plastic increased from 0.5% to 1.5%.

7.2.3.4 Effect of Normal Concrete

In M20 grade of concrete, the value of oxygen permeability was $18.18 \times$ E-16, which was similar when 1% plastic in 8 M of NaOH solution in ratio of Na_2SiO_3 to NaOH was 3 and ratio of SiO_2 to Na_2O was 3.35. For M25 grade of concrete oxygen permeability was $12.91 \times$ E-16, which was similar when 1.5% plastic in 12 M of NaOH solution in ratio of Na_2SiO_3 to NaOH was 3 and ratio of SiO_2 to Na_2O was 3.35 of geopolymer concrete.

7.2.4 CHLORIDE PENETRATION IN GPC WITH PCMPW

This test method covers the determination of the electrical conductance of concrete to provide a rapid indication of its resistance to the penetration of chloride ions. It measured the quality and properties of concrete. The test is described in ASTM C 1202.

The ability of concrete to resist the transportation of chlorides is an important factor in its potential durability. If chlorides can be prevented from reaching any steel in the concrete, then the risk of corrosion is reduced. This test method covers the determination of the electrical conductance of concrete to provide a rapid indication of its resistance to the penetration of chloride ions. The test method also provides an indirect measure of the permeability of the concrete, a critical parameter in all durability-related distress mechanisms. The lower the permeability, the longer the concrete will survive chemical and environmental attack. The permeability of concrete can be indirectly assessed by measuring the electrical conductance of a sample of concrete. Test Procedure – the test is described in ASTM C 1202. A 50-mm-thick section is obtained from a 102 mm diameter lab molded cylinder. The core section is completely saturated with water in a vacuum apparatus. Electrical current is passed from one side of the core section to the other side while it is contained within a cell that has a sodium chloride solution on one side of the core and a sodium hydroxide solution on the other side. The electric current is applied and measured for 6 hours.

Apparatus and material required for test – Vacuum saturation apparatus: completely saturates the sample, sodium chloride solution (3% by mass in distilled water), sodium hydroxide solution (0.3 N in distilled water); sealed cell: holds the core specimen with each liquid solution on opposite sides of the core section and has electrical leads for connecting a DC electrical source; DC power supply: provides constant DC power to the test specimen; voltmeter: measures and records volts and amps on both sides of the core specimen.

7.2.4.1 Test Method

- Completely saturates the core section with water.
- Covers exposed face of specimen with an impermeable material such as rubber or plastic sheeting. Place rubber stopper in cell filling the hole to restrict moisture movement. Allow sealant to cure per manufacturer's instructions.
- Place the saturated core section in the sealed cell containing the two different sodium solutions on either side of the core section.
- Connect the power supply and voltmeter.

- Apply a 60-volt DC current across the cell for six hours.
- Convert the ampere-seconds curve recorded from the test to coulombs.

Calculation: The current is recorded at 30-minute intervals; the following formula, based on the trapezoidal rule, can be used with an electronic calculator to perform the integration: $Q = 900 (I0 + 2I30 + 2I60 + ... + 2I300 + 2I330 + I360)$ where: Q = charge passed (coulombs), $I0$ = current (amperes) immediately after voltage is applied, and It = current (amperes) at t min after voltage is applied. If the specimen diameter is other than 3.75 in. (95 mm), the value for total charge passed established in 11.1 must be adjusted. The adjustment is made by multiplying the value established in 11.1 by the ratio of the cross-sectional areas of the standard and the actual specimens. That is, $Qs = Qx (95/X)2$, where Qs = charge passed (coulombs) through a 95-mm diameter specimen, Qx = charge passed (coulombs) through x in. diameter specimen, and X = diameter of the nonstandard specimen. Use Table 7.2 for evaluating the result.

7.2.4.2   Effect of NaOH

When the ratio of sodium silicate to sodium hydroxide was constant and the molar content of sodium hydroxide (NaOH) solution increased, results show the charge passed in coulombs was decreased, as the molar content of NaOH was increased then charge passed in coulombs decreased from 18.12% to 37.17% in SiO_2/Na_2O of 2.25 and 8.73% to 14.05% in SiO_2/Na_2O of 3.35.

7.2.4.3   Effect of Na₂SiO₃/NaOH

When the molar content of sodium hydroxide (NaOH) solution was constant and the ratio of sodium silicate to sodium hydroxide was increased, then the charge passed in coulombs of geopolymer concrete decreased. When the ratio of sodium silicate to sodium hydroxide increased from 1 to 3 then charge passed in coulombs of geopolymer concrete decreased from 0.50% to 11.95% in SiO_2/Na_2O of 2.25 and 0.25% to 3.93% in SiO_2/Na_2O of 3.35.

7.2.4.4   Effect of Metallized Polymer Plastic

For a constant molar content of sodium hydroxide, and constant ratio of sodium silicate to sodium hydroxide, the charge passed in coulombs of geopolymer concrete was increased with the increase of metallized polymer plastic. Charge passed in coulombs increased from 1.11% to 18.95% when SiO_2/Na_2O ratio was 2.25 and 1.14%

TABLE 7.2

Standard Reference Values for Chloride Penetration in Concrete

Charge Passed in Coulomb	Chloride Ion Penetrability
>4,000	High
2,000–4,000	Moderate
1,000–2,000	Low
100–1,000	Very low
<100	Negligible

to 37.17% when SiO_2/Na_2O ratio was 3.35 when percentage of plastic was increased from 0.5% to 1.5%.

7.2.4.5   Effect of Normal Concrete

In M20 and M25 grade of concrete, 862.21 and 616.99 coulombs were passed, which was nearly similar when 0.5% plastic in 16 M of NaOH solution and ratio of Na_2SiO_3 to NaOH was 3 and ratio of SiO_2 to Na_2O was 2.25 geopolymer concrete. When the ratio of SiO_2 to Na_2O was 3.35, the ratio of sodium silicate to sodium hydroxide increased as 1, 2, and 3; the molar content of NaOH increased as 8 M, 12 M, and 16 M; and the percentage of plastic increased as 0.5%, 1.0%, and 1.5%. The result shows moderate chloride ion penetrability of geopolymer concrete. When the ratio of SiO_2 to Na_2O was 2.25, the ratio of sodium silicate to sodium hydroxide increased as 1, 2, and 3; the molar content of NaOH increased as 8 M, 12 M, and 16 M; and the percentage of plastic increased as 0.5%, 1.0%, and 1.5%. The result shows low chloride ion penetrability, which was good for geopolymer concrete as compared to moderate chloride ion penetrability.

7.2.5   Accelerated Corrosion of Embedded Bars in GPC with PCMPW

The rebars, that is, the reinforcement steel bars, are conventionally added into the concrete. With time, the rebars develop corrosion and the oxides spread within the confining concrete mass. The corrosion spreads from the steel bars within the surrounding concrete as a function of the permeability of the hardened mass. As it has been observed that the GPC mixes have better capacities to resist the chemical disintegration due to the stronger molecular bonding, the addition of the waste plastic fibers should add the resistance further.

This test is based on electrochemical polarization principle. The experimental test setup essentially consists of a non-metallic container, in which water is mixed with 3.5% NaCl solution and filled to the required level. In this container, the cylindrical concrete specimen with rebar is placed centrally and around this a stainless steel plate is kept. The rebar of the concrete cylinder projecting at the top is connected to a DC power supply to the anode terminal (–ve) and the stainless steel plate to the cathode terminal (+ve). This setup forms an electrochemical cell with rebar acting as anode and stainless steel plate as cathode. Figure 7.20 shows a schematic view of the polarization test setup. A number of such units can be made and connected to a DC power pack of multi-channel system. A constant voltage of 5.0 V was applied from the DC power pack. Current is monitored with respect to time up to the propagation period. In this test the cylindrical specimens were casted 75 mm dia. and 150 mm length, which was embedded with steel bars of 87 mm length. After 7 days of casting, 24 specimens were cast, and the test was started and measured the weight loss of bar. The actual amount of corrosion of the bar is estimated by determining the gravimetric weight loss in the reinforcement bar, which was measured as the relative loss in weight over the length considered and represented the percentage weight loss. The sample test results are shown below and necessary detailed discussion is also presented.

The tests were conducted on the cylinder specimens with steel bars embedded in the center. The connections and arrangements are shown in Figure 7.19 and the

FIGURE 7.19 Accelerated corrosion test apparatus and specimens.

results are also presented to learn the role and function of the chemical proportions of the GPC making material and the plastic fibers.

The result shows the following observations were made and recorded to understand the responses by the GPC consisting of waste plastic fibers against the accelerated corrosive actions.

Current passed was 4–128 mA in 0.5% metallized polymer plastic, 1–162 mA in 1.0% metallized polymer plastic, and 3–370 mA in 1.5% metallized polymer plastic; corrosion initiation time was 16–25 days in 0.5% metallized polymer plastic, 16–23 days in 1.0% metallized polymer plastic, 14–18 days in 1.5% metallized polymer plastic; when ratio of SiO_2 to Na_2O was 2.25, the ratio of sodium silicate to sodium hydroxide increased as 1, 2, and 3, which increased the molar content of NaOH as 8 M, 12 M, and 16 M.

Current passed was 8–262 mA in 0.5% metallized polymer plastic, 7–438 mA in 1.0% metallized polymer plastic, and 12–576 mA in 1.5% metallized polymer plastic; corrosion initiation time was 14–18 days in 0.5% metallized polymer plastic, 12–15 days in 1.0% metallized polymer plastic, 8–13 days in 1.5% metallized polymer plastic. When ratio of SiO_2 to Na_2O was 3.35, the ratio of sodium silicate to sodium hydroxide increased as 1, 2, and 3, which increased the molar content of NaOH as 8 M, 12 M, and 16 M. In normal concrete current passed was 86–843 mA in M20 grade of concrete and 63–779 mA in M25 grade of concrete. Both grades of concrete have time for corrosion initiation, which was 22 days.

7.3 INFLUENCE OF PLASTIC WASTES ON THE PERFORMANCE OF CONCRETE COMPOSITE

The addition of plastic waste in fiber form showed several noticeable observations and outcomes from the experimental studies on the specimens consisting of conventional concrete and plastic fibers. The summary is presented as follows:

- The percentages of addition of metallized plastic waste increase and the workability of fresh concrete decreases. The maximum workability decreases in aspect ratio 1.5 as compared to the normal concrete workability.
- Addition of metallized plastic waste in concrete decreased the compressive strength. The percentages of addition of metallized plastic waste increased from 0% to 2%, compressive strength decreased from 0.60% to 24.18% as compared to normal concrete test result.
- Addition of metallized plastic waste in concrete decreased the split tensile strength in aspect ratio 1 and aspect ratio 1.5. The split tensile strength increased in aspect ratio 17.5 up to 1% of plastic waste.
- Impact strength also decreased due to the addition of plastic waste. If the aspect ratio increased, the difference between initial crack and final failure also increased. If the volumetric fraction of plastic waste increased, the difference between initial crack and final failure also increased.
- The percentage of plastic waste only affects the durability of concrete. From the observation of test results, there was no effect of aspect ratio on the durability of concrete. If the percentage of plastic waste was increasing, the durability was decreasing.
- The compressive strength of concrete in normal curing, sulfate curing, and acid curing was found to decrease when the metallized polymer plastic percentage increased. It is observed that when cubes were subjected to normal and sulfate curing the strength decreased in the range of 2%–30% but for cubes subjected to acid curing the strength was observed to be lower than normal curing in the range of 20%–30%.
- The tensile strength of concrete determined by split tensile strength and pull-off test was found to decrease with increase in metallized polymer plastic. The split tensile strength of concrete decreased in the range of 3%–46% and pull-off strength of concrete decreased in the range of 2%–50% as compared to normal.
- The oxygen permeability of plain concrete disk was observed to be 1.75–5.30 times higher than metallized polymer plastic plus fly ash concrete. Hence the concrete mix with 0% metallized polymer plastic plus 0% fly ash was observed to have minimum permeability.
- The weight difference: As the percentage of metallized plastic increased in the mix, the weight of cube subjected to acid curing was observed to increase in the range of 1%–4%, and in sulfate curing was observed to increase in the range of 0.1%–0.7%. Also, it was observed that when fly ash percentage in concrete mix increased from 0% to 10%, the weight of cube in acid curing reduced but when fly ash percentage increased from 10% to 30%, the weight of cube increased as compared to plain concrete.
- When the absorption of concrete disk was checked, at 300 seconds for addition of fly ash from 0% to 30% plus metallized polymer plastic from 0% to 1.5%, water absorption increased in the range of 20%–55% compared to plain concrete. But when the absorption of concrete disk was checked at 1,800 seconds it increased in the range of 30%–65% for addition of fly ash from 0% to 30% plus metallized polymer plastic from 0% to 1.5%.

7.4   INFLUENCE OF INDUSTRIAL WASTES ON THE PERFORMANCE OF CONCRETE COMPOSITE

This section deals with the preparation of concrete using industrial wastes, namely fly ash. The conventional binder of cement-based concrete was fully replaced with fly ash and alkaline solution to develop the binder matrix. Moreover, the addition of the other wastes namely metallized plastic has also been utilized as the fiber in the matrix. Following is the summary of how the wastes have influenced the properties of the concrete.

- The concentration of sodium hydroxide solid increased in sodium hydroxide solution from 8 to 16 M; it was observed that the compressive strength, split tensile strength, and impact resistance of geopolymer concrete were increased. And chloride ion penetrability decreased.
- For alkaline liquid to fly ash ratio, the compressive strength, the impact resistance, and the chloride ion penetrability for alkaline liquid to fly ash ratio of 0.45 gave better result as compared to the ratio of 0.55. And the split tensile strength for alkaline liquid to fly ash ratio of 0.55 was more as compared to 0.45 for constant sodium hydroxide concentration and sodium silicate to sodium hydroxide ratios of 2 and 3.
- For sodium silicate to sodium hydroxide ratio, the compressive strength, the split tensile strength, the impact resistance, and the chloride ion penetrability of geopolymer concrete for sodium silicate to sodium hydroxide ratio of 3 gave better result as compared to 2 for constant sodium hydroxide concentration and alkaline liquid to fly ash ratios of 0.45 and 0.55.
- In accelerated corrosion test, the actual mass loss of steel bar embedded in geopolymer concrete was negligible as compared to theoretical mass loss, but if the mass loss of steel bar is compared with different geopolymer concrete mixtures then as the sodium hydroxide solid concentration increased in solution, it was observed that mass loss decreased for constant alkaline liquid to fly ash ratios of 0.45 and 0.55 and sodium silicate to sodium hydroxide ratios of 2 and 3.

7.5   SUMMARY OF ASSESSMENT OF MODIFIED CONCRETE

This section gives the overall summary of development and assessment of concrete consisting of industrial wastes, namely fly ash and the plastic waste, in fiber form. All the salient features and conclusions have been summarized in the points given below:

- The molar content of sodium hydroxide increased from 8 to 16 M and the compressive strength (normal, acid, and sulfate curing) increased in geopolymer concrete.
- The molar content of sodium hydroxide increased from 8 to 16 M, while oxygen permeability, water absorption, weight difference (acid and sulfate curing), charge passed in coulombs (RCPT) decreased.

- For a constant molar content of sodium hydroxide, the compressive strength (normal, acid, and sulfate curing) of geopolymer concrete increased with the increase in the ratio of sodium silicate to sodium hydroxide as 1, 2, and 3.
- For a constant molar content of sodium hydroxide, the oxygen permeability, water absorption, weight difference (acid and sulfate curing), charge passed in coulombs (RCPT) decreased with the increase in the ratio of sodium silicate to sodium hydroxide as 1, 2, and 3.
- For a constant molar concentration of sodium hydroxide, ratio of sodium silicate to sodium hydroxide the compressive strength (normal, acid, and sulfate curing), oxygen permeability, Water absorption, weight difference (acid and sulfate curing). Charge passed in coulombs (RCPT) decreased and ratio of SiO_2 to Na_2O increased from 2.25 to 3.35.
- For a constant molar concentration of sodium hydroxide, ratio of sodium silicate to sodium hydroxide and SiO_2 to Na_2O ratio, compressive strength (normal, acid, and sulfate curing), oxygen permeability, water absorption, and weight difference (acid and sulfate curing) decreased with the increase of metallized polymer plastic as 0.5%, 1.0%, and 1.5%.
- In acid curing of 0.20%–1.0% and in sulfate curing of 0.24%–1.07%, there was weight difference during use of SiO_2 to Na_2O ratio of 2.25. While in SiO_2 to Na_2O ratio of 3.35 it was 1.03%–3.10% and 1.17%–3.49%.
- In rapid chloride ion penetration test, the ratio of sodium silicate to sodium hydroxide increased as 1, 2, and 3; the molar content of NaOH increased as 8 M, 12 M, and 16 M; and the percentage of plastic increased as 0.5%, 1.0%, and 1.5%. The result shows low chloride ion penetrability, which was good for geopolymer concrete as compared when the ratio of SiO_2 to Na_2O was 2.25, while in the ratio of 3.35 it shows moderate chloride ion penetrability, as per ASTM C 1202.
- Accelerated corrosion test shows the corrosion initiation time was 8–25 days. From this experimental study it was found to have good durability while using metallized plastic from 0.5% to 1.5% as compared to normal concrete.
- The number of fly ash-based geopolymer concrete containing metallized plastic increased the compressive strength, split tensile strength, impact resistance, pull-off strength, and rebound hammer, with the increase in the molar concentration of sodium hydroxide from 8 to 16 M.
- For a constant molar concentration of sodium hydroxide, the compressive strength, split tensile strength, impact resistance, pull-off strength, and rebound hammer increased with the number of geopolymer concrete containing metallized plastic with the increase in the ratio of sodium silicate to sodium hydroxide by mass.

  The compressive strength, split tensile strength, impact resistance, pull-off strength, and rebound hammer were maximum with the number of geopolymer concrete containing metallized polymer plastic having Na_2SiO_3 to NaOH ratio 3 and SiO_2 to Na_2O ratio 2.25.
- For a constant molar concentration of sodium hydroxide, ratio of sodium silicate to sodium hydroxide, and SiO_2 to Na_2O ratio, the compressive strength,

split tensile strength, impact resistance, pull-off strength, and rebound hammer decreased with the number of geopolymer concrete with the increase of metallized polymer plastic.

- Comparing strength properties between geopolymer concrete containing metallized polymer plastic and conventional cement concrete of grade M20 and M25, it was observed that geopolymer concrete containing metallized polymer was in the order of strength properties of conventional cement concrete.
- Impact resistance of the geopolymer concrete containing metallized polymer was observed much more than that of conventional cement concrete.
- All the strength properties of geopolymer concrete containing polymer plastic are comparable with conventional cement concrete, hence it can be used for construction.

8 Applications of Modified Concrete in the Civil Engineering Fields

8.1 UTILIZATION OF THE MODIFIED CONCRETE IN NON-STRUCTURAL APPLICATIONS

Construction activities may be divided into two broad categories as far as the loading or actions on the member are concerned. The first type is non-structural activities. These are activities where the concrete is not utilized for resisting the external loads or not subjected to structural actions such as compression or tensile forces. The examples are lean concrete used for leveling of the surfaces, backfilling or the secondary support to the main structures, lightly loaded walkways, barriers for traffic lanes, wall tiles, and surface panels. Though the concrete has to bear forces, the structural response is not expected.

We have explored three different types of concretes and have also examined them for their preliminary response to structural actions. However, the concrete unable to respond at the desired level may be useful for their non-structural usage. Following are some examples.

8.1.1 CEMENT-BASED CONCRETE MODIFIED WITH WASTE PLASTIC FIBERS

Concrete is sensitive and susceptible to cracking due to external and internal factors. Cracking is initiated with the very early stage of the settlement of the fresh mixes in the formworks. Therefore, to avoid cracking under different conditions, it is important to add the constituent that may help to reduce the crack formation. Over the past many decades, concrete has been modified with various types of macro and micro sized fibers of different material to support the cracking resistance and thereby improve the ductility and tensile strength resistance of the hardened mass. In this book, we have learned an innovative use of the fibers rendered from plastic wastes. The waste plastic belongs to the metalized plastic films used for food packaging and is one of the most hazardous wastes in the environment. Safe disposal of such plastics is a challenge and therefore, it would be wise enough to consider the addition of such wastes into the construction material as a secondary filler material. Following are some of the suggested applications of the modified concrete with plastic wastes.

- As macro fibers in cement-based concrete: The fibers are best suitable to be added in concrete for additional cracking resistance of the hardened mass. It has been experimentally confirmed that the use of plastic fibers improves

 DOI: 10.1201/9781032621340-8

the ductility of matrix to an extent. However, care should be taken that the length of the fibers may be restricted to 20 mm to avoid unwanted balling effects of the fresh mixes. In addition, an appropriate admixture should be used to balance the friction that may have increased due to the volumetric changes of the mass in the presence of the fibers.

- As aggregates in cement concrete: The author and his team have performed experiments for replacement of the natural aggregates by recycled plastic aggregates. The recycled plastic aggregates are lighter in weight. Therefore, they may be used in the development of lightweight concrete. The plastic aggregates are highly non-reactive to the chemicals and do not participate in the hydration process even at the granular size of 2 mm or less. Therefore, they can be utilized as an alternative for natural aggregates. However, the upper limit of the replacement should be maintained up to 30%–40% by weight of the aggregates. This is due to the limited strength of the recycled aggregates in resisting the compression. Moreover, the surface of the recycled aggregates is relatively smooth and may hinder the surface adhesion of the cement paste and therefore, the intermolecular bonding may get affected.
- As powder in the cement concrete: The recycled plastic in powder form may be replaced by up to 20% in concrete mixture. The powder form of the plastic remains unaffected by the chemical bonding as in case of cement with water. This is why the admixture of suitable adhesion capacity is very important in the concrete mixture. Though experimental works have been carried out on the topic, the strength in compression and specially splitting action reduces on increasing the plastic powder component in the mixture.
- In general, the concrete prepared with plastic wastes should be utilized in members that are not structurally loaded or expected to resist heavy loads during the service span. Another important aspect is that the plastics are susceptible to degradation in the presence of UV rays and show poor resistance to stability. Therefore, building components used for underground construction may be prepared with concrete consisting of plastic wastes in any form. In addition, the plastic wastes are not able to withstand fires. Therefore, the members or elements expected to have fire or high-temperature conditions should not be primarily made with concrete consisting of plastic wastes.
- Use of plastic waste in concrete is one of the ways to prepare a sustainable concrete. However, the suitable dosage, replacement of natural material, sufficient admixtures, and selection of application bear vital importance.

8.2 UTILIZATION OF THE MODIFIED CONCRETE IN STRUCTURAL APPLICATIONS

The concrete modified with plastic waste should be utilized in light loaded members or elements, for example, plinth beams, compound wall columns, concrete for mortar making in the foundation, damp proof course of the concrete on masonry, secondary elements of the buildings like lintel bends, plain concrete for leveling of the footing in the basement areas or members in the below-ground level, and up to an extent in the slabs not exposed to sunlight directly or permanently. The experimental work has demonstrated that the waste plastic in fiber form may be used in the structural concrete up to 1% of the volume of the concrete that resembles nearly 24 kg/m^3 of waste quantity.

The overuse of waste plastic in the structural elements restricts the load transfer and stress distribution mechanism. The plastic aggregates, plastic fibers, or plastic powder do not participate in the chemical process; therefore, they do not contribute to the strength gaining mechanism in the members. Moreover, in the case of recycled plastic waste aggregates, there are chances of having significant air voids and other entrapped air locations that on drying increase the cracking possibility in the mass. There is also a possibility of reduction in the weight of the mass in case of significant replacement of the aggregate from concrete. In this condition, the concrete should not be used in structural members as the minimum strength resistance expected from the dense concrete may not be obtained where the lightweight concrete due to the addition of the plastic waste is prepared and utilized.

Another type of waste generated in the industry from the various processes may be extensively used in the cement-based concrete or in the development of the new binder material, namely geopolymer concrete. In Chapters 6 and 7, it was established that the full replacement of the cement in concrete is possible and with the use of the alkaline solution, the composite may be prepared resembling concrete. The concrete prepared with fly ash and alkaline activators is also referred as geopolymer concrete. This is because the alkaline solution and the silica-rich fly ash develop a chain-like microstructural bond system due to the exothermic chemical process called polymerization. The chemical process or action takes place when the mass of the concrete is provided with the extra heat or temperature by means of oven or heat curing. The process of development of the concrete is little challenging and requires experience of several mixes and trials. However, the material shows excellent strength and durability resistance.

The concrete prepared with fly ash may be more environmentally friendly compared to cement-based concrete, as it does not use cement as the binder material. Following are the points to be taken care in case of making geopolymer concrete:

- The scale and construction method: If the members are of larger dimensions, namely 1.2 m in a linear dimension, the geopolymer concrete should be prepared using fly ash, silica fume, and slags. The addition of silica fume is important to generate internal heat in the matrix to initiate the polymerization process for the bond formation. The conventional method of making geopolymer is based on the external heating method. This method is suitable for small and medium-scale members.
- Those linear members or planar members that are difficult to be heat-cured are to be provided with the steam curing option. The controlled heating of the member cast into the form work or molds should be exposed to the steam generated around the members in a closed and protected arrangement. This method has the limitation of uncontrolled heating. The heating should be continuous and temperature should be maintained at a constant rate, which is difficult to obtain in steam curing conditions.

- The larger sized members should be prepared with the wastes including fly ash, blast furnace slag, and silica fume mixed in a dry form together and added with the alkaline solution. This is called one-part geopolymer concrete. It is the trending technology for the manufacture of geopolymer concrete without external heat supply. This is more suitable for making the members of any scale since the heat curing is not a constraint. Moreover, with appropriate choice of admixture or the plasticizers, the fresh behavior of the concrete can be effectively managed and the material may be used for in situ casting of members at the construction sites similar to conventional concrete. However, the geopolymer produces more heat of hydration and therefore, care should be taken to minimize the damage due to the surface and inner mass access cracking. The one-part geopolymer concrete is gaining momentum and several research works may be referred to the same addressing fresh, hardened, and durability responses of the material and specimens prepared thereof.

8.3   UTILIZATION OF THE MODIFIED CONCRETE IN PRE-CAST CONCRETE MANUFACTURING

Pre-cast concrete technology has been claimed as the construction method of the 21st century. The pre-cast concrete products are extensively being utilized nowadays in practice and are very well accepted by professionals. There are several salient features associated with the technology. Some are discussed as follows:

- One of the major advantages of the pre-cast concrete is the quality of the material and product. The pre-cast members prepared with conventional concrete also respond better than the cast in situ conditions. This is true because, from mixing to curing, all stages are within the monitoring range and highest control conditions.
- Pre-cast members are convenient to transport within the manufacturing units and site routs. This adds the benefit of avoiding unnecessary storage of raw materials at the sites. Moreover, the waste of material can be considerably reduced. At the workshops or manufacturing units, excellent controls for material usage and mix design become possible, which further increase the quality of concrete and members.
- The pre-cast members are easier to connect and do not need time for settlement of the material. This ultimately results in faster rate of construction as well as faster completion of the units with less time.

Though pre-cast concrete has shown several advantages over conventional methods, very large pre-cast members need attention for shifting and transporting.

As far as the use of modified concrete, namely concrete with plastic waste and concrete prepared as geopolymer concrete, is concerned, for application with pre-cast technology methods, point required to be focused and to be taken care is:

The mix design should be balanced in such a way that the minimum concrete strength as per the pre-stressing forces is fulfilled.

8.4   UTILIZATION OF THE MODIFIED CONCRETE IN ASPHALT-BASED CONCRETE

We have discussed applications of wastes from communities and industry in preparing cement-based concrete or cement-based mortars extensively in the earlier sections. Another area where the waste plastics and industrial by-products may be utilized is the development of asphalt-based concrete in flexible pavements.

We know that bituminous concrete or asphalt-based concrete is a common material for use on flexible surfaces on state or national highways. Considering the preliminary mix formation, the plastic waste may very well get mixed in the concrete. In fact, there have been examples wherein the plastic bags and similar LDPE-based wastes are utilized in recycled form for road constructions.

The bitumen-based mortar is highly stiff in nature unlike concrete prepared with cement. The elastic properties and stability of the freshly mixed asphaltic concrete need attention before getting added with the plastic wastes or other pozzolanic industrial wastes such as fly ash, silica fumes, and furnace slag. The mixture of bitumen-based concrete should show significant workability and also provide excellent adhesion on drying with the aggregates. When such mixes are being modified with the wastes, the optimum amount of the addition should be carefully determined by employing necessary pilot studies.

There have been specific tests mentioned in the standards, namely

- Kinematic viscosity
- Shear test
- Flash point
- Binder content test
- Characterization of the mixture

Above are a few of the important laboratory investigations and tests to qualify the asphalt mixtures or bituminous concrete. A few recent studies have contributed with many important attributes in this area. For example, researchers have commented on the use of waste aggregates obtained from concrete recycled from the asphalt concrete (Ahmad et al. 2023). The team has performed very interesting tests on the modified mixes for moisture retention aspects. It is well known that the recycled aggregates possess larger capacity for moisture retention and the same should be considered while using them in concrete for pavements. The tests showed that up to 30% of the addition of the waste concrete aggregates may be an acceptable replacement ratio including the increase of the bitumen content up to 5.7% in the mix. This shows that the addition of the waste concrete in bitumen-based concrete needs additional binder content. However, on the other hand, use of such aggregates increased the tensile strength of the mixture significantly by up to 80% compared to the control mixture.

Other wastes namely plastic and used oils are utilized in the making of asphalt concrete. A team worked on such experimental investigations using 18% of the waste plastic by weight of the bituminous content in the mixture and varying proportions of the waste engine oils, up to 40% (Ogada et al. 2023). The results have been encouraging and in support of such applications. In another similar work, the authors have demonstrated possible use of the recycled municipal solid waste and fly ash in asphalt-based concrete (Joumblat et al. 2023). Fly ash has been used as partial to full replacement for filler material in the concrete. The test results showed increase in the rutting resistance and cracks due to fatigue effects on the flexible road surfaces prepared with modified asphalt concrete using fly ash and waste materials. Other materials namely waste glass have been utilized in the flexible pavements as partial replacement in asphalt concrete (Peng et al. 2023). The characteristics of the pavements, namely thermal resistance, increased and supported in reducing the heat-island effect at a high environmental temperature condition. There are several examples available wherein the asphaltic concrete is modified with plastic wastes and other industrial waste materials and has shown promising results. The following are some of the advantages to support such modifications:

- Using plastic wastes in asphalt concrete may be an environmentally safe and better waste disposal option compared to conventional methods.
- Modification of the binder bituminous mixture in the asphalt concrete with pozzolanic fillers like fly ash and slag may be supportive to a stable mixture of the concrete.
- Common challenges for flexible pavements such as fatigue and surface wear may be reduced by incorporating waste tires and similar waste products.
- Surface cracking due to temperature variation may also get controlled and mitigated with the use of recycled plastic and recycled concrete with an optimum dosage addition in the mixture.

Along with the benefits, there are a few limitations as well for the use of wastes in asphaltic concrete. Some of them are mentioned below for a general discussion; however, a detailed experimental study may lead to an exact understanding of such applications:

- In case of replacement of the aggregates of the mixture, with the waste or recycled concrete aggregates, optimum or maximum replacement ratio restricts the extensive use of the waste.
- The presence of fillers like pozzolanic materials may lead to changes in the minimum kinematic viscosity of the binder bituminous material. The excessive addition of the minerals in such concrete may create stiff and non-flowing mixtures that may result in inadequate binding of the constituents.
- Liquid wastes such as molten waste plastics and used oils may not be utilized along with the binder material as the cohesion and adherence of the other constituents may not help in achieving sufficient bonding.

The area and the avenues pertaining to the use of wastes in asphaltic concrete are promising areas of current research. Overall, waste in many forms has been added in concrete and has shown potential for their better utilization and similarly mitigation of pollution.

8.5   REFERENCES OF THE NATIONAL AND INTERNATIONAL CODES OF PRACTICES

Waste utilization in concrete has remained one of the most popular and ever-emerging areas for researchers. However, there are a few standard codes dealing with such applications in practice. As far as concrete prepared with cement is concerned, there are no guidelines dedicatedly prepared. Nevertheless, the Indian standard codes have mentioned the use of a few most commonly generating industrial wastes as discussed below:

- IS: 3812-1-2013 is an Indian code about the use of fly ash in concrete and cement mortars. The code defines many types of the ashes, namely bottom fly ash, mound fly ash, calcareous fly ash, siliceous fly ash, pond ash, and pulverized ash generated during the different thermal power generation processes. The method of collection of the fly ash is different and based on chemical compositions and the fuel used in the power stations; the ashes differ from one another. Therefore, they are utilized in various ways and also in the production of different composites. It is necessary to refer to the chemical and microstructural properties of the given type of ash before use. The present code also specifies the test types and the required code to perform chemical analysis of the fly ash samples.
- Along with the chemical properties, the ash may be evaluated for physical properties also, namely fineness, particle size, lime content, soundness, and strength of the material in compression. One more aspect that is covered in the code is knowledge about detecting impurities in the sample. The ashes consist of a fine powder and are generated from a combination of several particulate matters. The purity of the material is important and must be confirmed by the supplier or manufacturer.
- IS: 10262-2019 is an Indian code that deals with the mix design of concrete prepared with cement as the binder material. The code indicates inclusion of fly ash of a given grade to be used in the mix design process and calculations. This shows that fly ash is quite a commonly used waste in construction. The code describes step-wise methodology for preparing the theoretical quantity calculations of all the major ingredients, such as cement, sand, aggregates, water, and admixtures. Depending upon the strength or the target strength of the mixture, pozzolanic materials such as fly ash may be utilized as mentioned in the code. It is to be noted that the code indicates use of fly ash up to a certain percentage of replacement of the cement. The total content of the binder material comprises cement and fly ash wherein fly ash may be replaced by up to 40% as the current upper limit of replacement to meet requirements of the minimum cement content in the given mix.
- IRC-121-2017 is an Indian code that specifies the use of the manufactured sand commonly obtained from construction and demolition wastes as the replacement of natural sand for road pavements and similar construction activities. It is an informative code regarding the utilization of the construction and demolition residues of varying nature, namely the hardened plaster

mortar concrete, reinforced concrete, crushed bricks, and mixed nature waste and many others. The code also indicates the tests and results carried out by international organizations and researchers on the topic and provides a summary of construction and demolition waste usage in the construction for pavements. The document also includes minimum acceptance ratio of the properties of the aggregates obtained from the pre-processing of the demolition wastes such as aggregates and sand. However, researchers or the professionals are required to carry out the necessary tests on each of the materials obtained from the demolition wastes before its addition in any conventional composite or matrix. The author has published an article on the pilot studies and results obtained regarding the use of waste concrete for its application in making concrete for construction activities, which may be of interest to the readers (Modi and Bhogayata 2023).

- Another interesting reference in the form of a special publication by the Indian Road Congress standards has been made available as guidelines for using plastic wastes in asphaltic concrete in the top layer of flexible pavements. This is one of the ways of using waste hazardous plastic in relatively less structurally important applications in the construction field such as the asphalt concrete is used for the preparation of the top layer of the flexible pavements. The guideline suggests use of the waste plastic in shredded form. The specific size of the pre-processed plastic granules or flakes has been mentioned for reference. However, the design is a guideline, and there are other features, namely method of mixing, cleaning of the waste plastic, and use of chemicals, to maintain the preliminary properties of the asphalt concrete that require attention and also to be checked by employing pilot studies.

- Internationally, the ASTM standard on making cement-based building blocks may be referred. ASTM C55-11 guidelines are available for reference wherein the different waste materials have been mentioned, namely slag and silica wastes from the industrial activities may be added into the mixture.

- The American codes namely ACI-211.1.-91 provide guidelines on how to utilize silica fumes, slag, and fly ash in the production of mass concrete for construction. However, the readers are advised to perform pilot studies like any other research work before making the final decisions on the mix designs.

- In summary, there are various national and international documents in the form of codes or guidelines that are available for reference; however, the exact design methods are not explicitly mentioned as a standard document till date.

REFERENCES

Ahmad, J., et al. "Investigation on the effect of moisture induced damage on asphaltic concrete mix incorporating waste concrete aggregates." *IOP Conference Series: Earth and Environmental Science*. Vol. 1151. No. 1. IOP Publishing, 2023.

Ogada, Joyce Susan Liavuli, Sixtus Kinyua Mwea, and George Matheri. "Performance of plastic waste and waste engine oil as partial replacements of bituminous asphalt concrete in flexible pavement." *East African Journal of Engineering* 6.1 (2023): 48–65.

Joumblat, Rouba A., et al. "Investigation of using municipal solid waste incineration fly ash as alternative aggregates replacement in hot mix asphalt." *Road Materials and Pavement Design* 24.5 (2023): 1290–1309.

Peng, Bo, et al. "Semi-flexible pavement with glass for alleviating the heat island effect." *Construction and Building Materials* 367 (2023): 130275.

IS: 3812-1-2013 (https://law.resource.org/pub/in/bis/S03/is.3812.1.2013.pdf)

IS: 10262-2019 (https://law.resource.org/pub/in/bis/S03/is.10262.2009.pdf)

IRC-121-2017 (https://law.resource.org/pub/in/bis/irc/irc.gov.in.121.2017.pdf)

Modi, Ravi, and Ankur C. Bhogayata. "Utilization of recycled concrete residues as secondary materials in the development of sustainable concrete composite." Materials Today: Proceedings (2023). (Article in press) https://doi.org/10.1016/j.matpr.2023.03.664

IRC: SP:098-2013 (https://law.resource.org/pub/in/bis/irc/irc.gov.in.sp.098.2013.pdf)

ASTM C55-11 (https://www.astm.org/c0055-11.html)

9 Life Cycle Assessment and Economic Advantages of Modified Concretes

9.1 LCA OF CONVENTIONAL CONCRETE

There are several literary references and documents available on LCA for conventional concrete. The topic has emerged as one of the major attractions by researchers in the past few decades. It is well known that LCA is a comprehensive approach to assessing the environmental impacts of a product or service, taking into account all stages of its life cycle. It is based on the cradle-to-grave analysis and includes all the inputs and outputs associated with the product or service, such as energy use, material consumption, waste generation, and emissions to air, water, and land. The aim of LCA is to provide a complete picture of the environmental impact of a product or service, which can help inform decisions about product design, material selection, process optimization, and waste management. By identifying the areas of greatest environmental impact, LCA can help guide efforts to reduce the overall environmental impact of a product or service throughout its life cycle. LCA is increasingly being used by companies, governments, and other organizations as a tool to support sustainability and environmental management initiatives.

The primarily important aspects of LCA are to define a goal and scope of the analysis. This is necessary because in general LCA may be conducted using several parameters and factors. In this regard, the goal of the present LCA is to quantitatively analyze the concrete-making process for energy consumption and the production of the associated pollution types. The scope of the LCA is to discuss qualitative analysis and identification of the stages requiring energy and release of pollutants. The boundary conditions are also necessary to be mentioned, which are the limitations of the LCA. As in the present case the LCA will not include the environmental influence of the concrete production at the micro level.

9.1.1 LCA FOR CONCRETE

The goal of LCA is the generation of information regarding the three major aspects of concrete, namely raw material processing, energy consumption during various stages and processes, and the emission of pollutants during the complete cycle of concrete usage. The complete cycle of usage includes raw material procurement and

DOI: 10.1201/9781032621340-9

production, mixing and proportioning, casting or laying of concrete, and services of functions intended from concrete during its hardened state.

9.1.2 LCA Attributes of Conventional Concrete

9.1.2.1 LCA for Materials and Ingredients

Cement is one of the key factors affecting LCA process. Cement is an energy-intensive material and therefore requires detailed evaluation. The following are the important stages for the cement making process and associated LCA attributes (Table 9.1).

From the general cement making process, it may be observed that the entire process of obtaining cement from a raw material is highly energy intensive. The manufacturing not only demands high amount of energy but also produces wastes such as dust, ashes, and sludge to an extent. More detailed analysis may be carried out with the help of the quantitative data collection. As a rough estimation, 1 ton of cement production emits 1 ton of carbon dioxide or similar gases in the air during the processing.

The kiln within which the fusion and formation of the cement powder take place is another source of energy demand and where the efficiency is of primary concern. The high burning temperature of nearly 1,500°C or more is needed for limestone to get converted into clinkers. This process also produces wastes such as ash and air particulate. All such activities may be separately investigated for LCA and more detailed results may be obtained. Ultimately, the objective is to understand the energy needs of a process and the pollution generated during these processes.

As explained, there may be qualitative and quantitative LCA employed for all materials. The LCA may provide the following valuable information for a specific ingredient of concrete:

- Type and nature of the fuel required for stage-wise process
- Efficiency of the manufacturing system

TABLE 9.1
Cement Making Process and Relevant LCA Attributes

Process

Mining of limestones	Fuel, machine operations, transportation	Air pollution, depletion of natural resources
Limestone processing for cleaning, grinding, and storage	Fuel, machine operations	Air pollution by fuel burning and particulates in air due to grinding and cleaning
Clinker making by burning and fusion of the limestone	Fuel for kiln machine operations	Dusting, air pollution, solid waste generation, ash formation
Fine grinding of clinkers and powder making	Fuel, machine operations	Dusting and air pollution
Packaging and logistics	Fuel, machine operations, transportation	Air pollution

- Material usage scenario
- Details of pollution emission and energy demands for overall system
- Nature of pollution, namely land, air and water, during the process
- Quantum of pollution and energy demand ratio
- Hazards to the environment
- Areas of improvement of the system of manufacturing/processing
- Socioeconomic impacts of making of the specific ingredient
- Research and innovation areas and topics for sustainability
- Possibility of incorporation of new methods and processes
- Attainment level or success ratio of energy vs pollution
- Rate of depletion of natural materials
- Information about associated climate changes

Above are samples of the information that may become available by conducting LCA. However, for conventional concrete, a detailed study is required as the LCA incudes the cradle-to-grave and cradle-to-gate approaches. Both the approaches may be employed as per need. The cradle-to-grave model discusses the generation, usage, and disposal of the material till the end of the useful life of the material or ingredient, while the cradle-to-gate model discusses the material up to the application stage in general. The service life response and the final disposal stage of a material can also be studied separately. This part includes the sustainability aspects. Once the material is in place and in use, the average life span, the functional details, the deterioration and final disposal, the waste generated, and its environmental impacts may be obtained with better clarity. Figure 9.1 shows the LCA of conventional concrete illustratively.

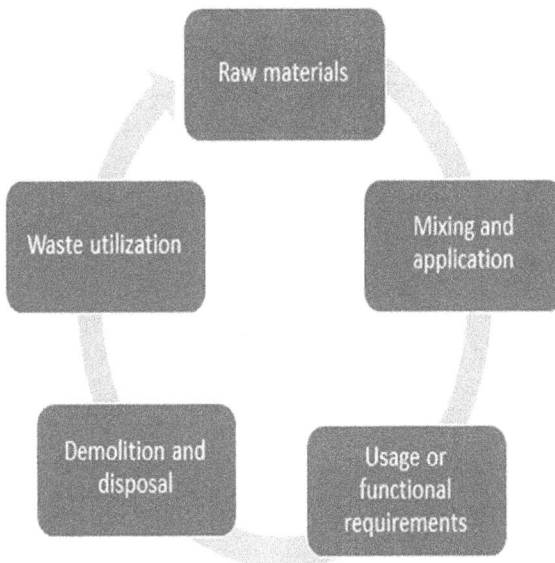

FIGURE 9.1 Generalized approach for LCA of conventional concrete.

9.2 LCA OF CONCRETE CONTAINING PLASTIC WASTES

The life cycle assessment (LCA) of concrete containing plastic waste involves analyzing the environmental impacts of the entire life cycle of the material, from extraction of raw materials to end-of-life disposal. The analysis considers factors such as energy use, greenhouse gas emissions, water consumption, and waste generation.

The use of plastic waste in concrete can reduce the amount of virgin materials needed for concrete production, thereby reducing the environmental impacts associated with the extraction and processing of these materials. Additionally, incorporating plastic waste can reduce the amount of waste going to landfills, which can also have environmental benefits.

However, the use of plastic waste in concrete can also have potential drawbacks. For example, if the plastic is not properly sorted and cleaned, it may contaminate the concrete and affect its properties, leading to durability issues and potentially requiring more frequent maintenance or replacement. Furthermore, the energy required to transport and process the plastic waste may offset some of the environmental benefits of its use in concrete.

The LCA of concrete containing plastic waste requires a careful consideration of the potential environmental benefits and drawbacks, as well as the economic and practical feasibility of incorporating plastic waste into concrete production. In the following sections, the advantages and limitations of using plastic wastes in concrete are presented.

9.2.1 Benefits of Using Waste Plastic in Concrete from the LCA Perspective

In Chapter 5, we noted that the use of waste plastic in concrete is advantageous as it addresses the utilization of the waste with a significant quantity. However, from the LCA perspective, it is important to review such modifications of a conventional material. Let us examine the LCA of the modified concrete with plastic wastes.

9.2.1.1 Goal and Scope of LCA

The preliminary goal of conducting LCA of modified concrete is to evaluate the sustainability and environmental impacts of the use of waste plastic in concrete. Cater the goal, a systematic assessment and evaluation should be performed considering all important stages of making of the composite. In addition, the boundary conditions should be employed to limit the application of the LCA. This means, there should be a clear start and end point defined for the end user for its effective use. What is included and excluded needs to be specified in all respects.

9.2.1.2 Goal of LCA

1. Study of the energy requirements of the concrete-making raw materials.
2. Obtain the influential parameters related to the procurement of the raw materials.
3. Understand the impact of use of natural material in concrete on environments.
4. Production scenario of plastic waste.

5. Processing of the plastic waste and conversion of the same into the useful ingredient for concrete.
6. Quantification of the plastic waste to be used in a unit volume of concrete.
7. Quantification of the natural ingredients possible to replace with waste plastic in varying forms, namely aggregate or sand.
8. Identify the influence of the addition of plastic waste on the concrete properties.
9. Obtain the energy consumption and saving for reduced usage of the aggregates and sand by waste plastic.
10. Qualitative analysis of direct and indirect advantages using plastic waste as one of the ingredients in concrete by replacing natural materials.

9.2.1.3 Scope of LCA
1. The supply of raw materials will be consistent from a single source and constant throughout the study. The geographic location will remain the same for the material procurement.
2. The transportation and other logistics will be considered as per the prevailing commercial values. However, the cost analysis will not be focused at any stage.
3. The influence of emission of the greenhouse gases will be considered from raw materials, processing of waste plastic, and mixture.

9.2.2 LCA AND ITS PURPOSE

LCA may be observed as a systematic approach of understanding the energy requirements, use of raw materials, and the service life and afterlife of the concrete consisting of the waste of different categories including plastic wastes. In this section the focus is on the plastic waste used in the concrete. The following are the guidelines for LCA on such modified concrete.

9.2.2.1 Inventory Analysis
This is one of the preliminary but most important attributes of the LCA process. The inventory analysis deals with the generation of the quantitative data necessary to develop the consecutive database. To hold the LCA on concrete consisting of plastic wastes, the target quantity to be created should be identified. Let us assume that we need to produce a $1\,m^3$ quantity of concrete with plastic wastes. The next step is to obtain the mix design. The mix design indicates quantities of the ingredients namely cement, aggregates, sand, and water along with the admixtures. Once the conventional quantities are obtained, the waste plastic quantities are required to be determined. Here, the replacement ratio plays an important role. The replacement should be done on the basis of the pilot studies and tests necessary to qualify the concrete for the intended applications, namely structural or non-structural applications. There are several options available to replace the conventional ingredients with plastic waste as we have seen in the earlier sections also, the specific raw material may be recalculated in accordance with mix design calculations and addition of waste plastics. The general approach is to replace the fine or coarse aggregates with the recycled plastic

granules, and using shredded plastic waste into the fiber form also. Here, the fibers are added into the concrete by the fractions of the total volume of the mixture.

The purpose of quantification of the ingredients is to obtain the energy demand by the raw material at various levels. The next step of inventory analysis is to quantify the total energy required for obtaining the raw material from the natural location or from the source to the process point. In our case, let us examine the stage of the obtaining aggregates for instance. Table 9.2 indicates the features to be included in this inventory.

From Table 9.2, it can be assumed that at nearly each stage there is a significant energy demand in obtaining the crushed rock base aggregates. The inventory can be extended for all other materials and especially for cement since it is considered as one of the most energy-intensive processes in concrete making. If all materials are explored for their stage-wise energy demand the total energy may be obtained in producing 1 m^3 concrete. As an estimate, to produce a meter cube of normal concrete the approximate energy required is 2,775 MJ and that again varies with the increase in the cement portion in the mix. On the other hand, for inventory analysis estimation of the working life energy or the embedded energy and the energy required to fully dispose of the used concrete required to be addressed is needed. This is the second phase of the inventory analysis wherein the stages of mechanical demolition, separation, sorting, cleaning, crushing, grinding, pulverizing, waste collection, and finally waste disposal to landfill should be considered.

If we look at the waste inclusion such as plastic in concrete, the process again requires the third stage of inventory analysis for waste plastic, namely waste collection, sorting, transporting, shredding, and transforming the waste into the mixture, which are all energy-intensive stages. This means, none of the processes included in the concrete prepared with the plastic waste is free of the energy demand and therefore, a careful calculation of the energy requirement and thus the inventory analysis should be employed.

TABLE 9.2

Inventory Analysis for Aggregates to Prepare Concrete

Inventory Item	Stage	Processes	Energy Demand Approximation	Relevant Environmental Concerns
Aggregate for concrete preparation	First stage of the system	Mining/digging/ excavation	0.45–0.50 GJ/ton	Particulate matter in air, loss of minerals, loss of land area
	Intermediate stage	Rock cutting		Vibration of soil, vanishing of natural resources
	Intermediate stage	Hauling		Air pollution due to fuel burning
	Intermediate stage	Crushing into smaller sizes	0.12–0.20 GJ/ton	Land pollution due to non-utilized waste
	Intermediate stage	Loading into vehicles		Air pollution due to fuel burning
	Final destination	Transportation to a given destination		Air pollution due to fuel burning

9.2.2.2 Impact Assessment

After completing meticulous estimation for the energy demand, the next stage of LCA for concrete with plastic waste may be impact analysis. This stage deals with the effects of material making on the environment to a large extent. As we have already seen, energy is involved at each stage of the concrete-making process, and the process also results in environmental impact. Generally, the impact is taken as an adverse effect only. However, this may not be true for concrete containing plastic waste. The stage-wise associated impact of the concrete prepared with the waste plastic is mentioned in the following subsection. In this assessment, the positive and negative sides are to be considered simultaneously for the material. Even though the production of the material pollutes the environment, its waste is also being utilized; hence, a positive assessment also exists.

The impact of preparing concrete should be quantified using the existing environmental parameters and standards. For example, to produce 1 metric ton of concrete mix, how much water is required? Answering this will provide an idea of the total water consumption by the concrete and to what extent the concrete will be impacting on water demand. Similarly, the amount of fuel required for transporting 1 metric ton of the concrete between two places will answer the question of impact caused by the air pollution generated. Likewise, all the major and minor necessary processes may be evaluated for their concerned pollution emitting events.

9.2.2.3 Interpretation

The LCA of a material should be analyzed and prepared within several contexts including the energy demands, use of raw materials, resource availability, vanishing rate of the material, and pollution resulting from each minor and major processes. The LCA results are important evidences of the usefulness of the material to mankind and nature.

The interpretation of the result of the LCA for concrete prepared with the plastic waste may be completed using the below self-assessment questions:

- How much quantity of concrete was considered in the study?
- What conventional materials were replaced with the plastic waste?
- What were the quantities of the concrete-making material and plastic waste considered in the study?
- What amount of the energy was calculated for fully developed modified concrete?
- What were the significant pollution types observed getting emitted during the manufacturing of the modified concrete?
- Which processes are most hazardous to the environment and which ones are friendly?
- What are the disposal requirements of the concrete prepared with the plastic wastes?
- What additional energy and pollution attributes are associated with plastic waste being used in the concrete and how can it be further mitigated?

Answering the above assessment questions leads us to the effectiveness of the LCA conducted on concrete prepared with the plastic wastes. Though the scope of this

section is limited and it may not be possible to carry out detailed LCA of the modified concrete, the readers may get a starting point for the study on the topic with
reference to the information shared here and may be useful for future studies.

9.3   LCA OF CONCRETE CONTAINING INDUSTRIAL WASTES

9.3.1   Industrial Wastes

Fly ash, furnace slag, silica fumes, municipal waste ash, bottom and pond ash, mining wastes, used foundry sand, and agricultural residues are all being considered as
industrial wastes nowadays as alternative materials for addition in concrete. The conventional concrete consists of three phases namely binders, inert and water or liquid
additives. The LCA should be exercised keeping the following attributes in focus for
industrial wastes added into conventional concrete.

- Grade and mix design of concrete
- Resources of the conventional and waste materials
- Application of the modified concrete or the purpose of manufacturing
- Major and minor processes involved in manufacturing
- Energy needs of each stage or process involved or identified
- Aim, scope, and goal of the LCA for modified concrete
- Current practices of making, using, and demolition of such concrete
- Afterlife disposal aspects for demolished concrete
- Environmental impacts by use of raw materials, processes, and demolished
  concrete
- Service span of the modified concrete and deterioration scenario
- Durability aspects of the modified concrete

Above are the attributes or aspects providing support in making a framework for the
LCA process for the intended material. Since we already have seen a few essential
aspects of LCA on concrete, the same are not repeated but a framework is presented
that is useful in generating the LCA analysis for the concrete containing wastes mentioned in the beginning of the discussion in this subsection.

The LCA for concrete consisting of industrial wastes may be planned as per the
below-mentioned framework:

- Define the aim of LCA (any or all may be taken into consideration)
  - For cost comparison or economical attributes
  - For environmental concerns
  - For production and manufacturing efficiency
  - For material consumption attributes
  - For revealing the impacts on socioeconomic relationship
  - For structural or engineering performance and efficiency
  - For possible refinement of the existing manufacturing system
  - For obtaining carbon footprints of the entire process
  - For total energy demand
  - For cradle-to-grave model or cradle-to-gate model

- Define the scope of the LCA (all should be considered for a wider scope)
  - The raw materials will be quantified for specific concrete quantity.
  - The mix design will govern the quantity calculations.
  - The energy consumption will be considered in joules.
  - The industrial wastes will be included in energy consumption.
  - The replacement ratio will have an upper cape limitation.
  - The water consumption should be considered qualitatively.
  - The LCA will consider mixing, transporting, and placing stages only.
  - The LCA will exclude cost comparison of raw and waste materials.
  - Socioeconomic aspects will not be covered in the LCA.
  - Carbon footprints will be focused in the mixing and placement stages.
  - Emission of greenhouse gases will be focused at all three stages.
- Impact assessment
  - The greenhouse effect will be considered.
  - The raw material consumption will be considered.
  - The electricity, fuel, and manpower quantification will be included.
  - The loss of minerals and earth material will be included.
  - The end of life or service life will be calculated.
  - The rate of deterioration of the concrete structures will be considered.
  - The recycling or pre-processing of industrial waste will be counted.
- Inventory analysis
  - All raw and waste materials will be quantified.
  - The energy needs by each constituent of concrete will be calculated.
  - The input and output parameters will be considered.
  - The material flow analysis will be considered for each constituent.
  - The total material quantity, total energy demand, and total output will be calculated for the given mixture of concrete consisting of industrial wastes.
  - The material durability aspects and performance may be included.
- Interpretation
  - The data collection, data processing, and data evaluation will be based on the cradle-to-grave model.
  - The model will be based on unit input and unit output scale.
  - The final outcome will be streamlined as quantities of total material procured or obtained, total energy consumed, and total environmental impact generated at the making of fresh mixture of concrete.
- Reporting
  - The document will consist of all inventories for natural and waste materials.
  - The report will specify all the stages of LCA with flow charts or diagrams wherever necessary.
  - The LCA will be conducted by collecting databases and actual collection of the information. All databases will be shared with the references and resources for transparency and authenticity.
  - The results and reports will be utilized for the intended purpose only.

Above are the typical framework suggested to the readers for developing the LCA of the concrete consisting of industrial waste materials. The same may be used for the general purpose concrete also. The readers may add or modify the framework according to the actual requirements.

9.4 ECONOMICAL ASPECTS, COST COMPARISON, AND LOGISTICS OF MODIFIED CONCRETE

The process of LCA may or may not include economic aspects of the material preparation and manufacturing. The reason is the scope of the LCA. In general, the LCA of concrete may be prepared for environmental aspects. However, the system offers scope to work on the related economic and other aspects also. In this subsection, a brief discussion is presented on the economic aspects of the making of modified concrete.

The economic attribute of concrete deals with the cost per unit volume of the material. The cost therefore should represent the circular economy or should start and end at the common point. For concrete, the following are the economic aspects required to be considered:

Cost of raw materials from source to destination: This should include all the possible expenses namely land acquired for mining on a lease from the authorities, total quantity being extracted from the mines in a given timeframe, the machine and manpower expenses, the running cost of daily basic activities, the transportation of raw materials being lifted, and the other expenses such as electricity.

Generally, the selling price of a material in a nutshell should be obtained by considering other hidden expenses, namely labor cost, taxes, duties on selling the goods, and importantly the storage and staking of the material at the specific location and that too for the given time interval. It is to be noted that the storage of material increases the overall cost significantly. On the other hand, long-term storage may result in a decrease of the properties of the materials due to environmental conditions, for example, cement is one such material that is sensitive to environmental conditions.

The concrete prepared with the conventional material and wastes should be evaluated for overall cost by considering the expenses on raw natural materials and on the other hand savings by reduction in their quantities due to the use of wastes. In fact, all the concrete-making material will demand some expenses whether they are waste materials or not. Therefore, the cost calculation should be realistic as far as possible.

Transportation or the logistics is an important parameter for obtaining the right economic equations on the concrete-making process. Nowadays with the common use of the ready mixed concrete, this component plays an important role. The concrete being transported may cost more compared to the concrete being prepared at the site. For large sites, it is advisable to install a temporary mini concrete-making plant by staking of the materials in a sufficient quantity. This will further bring the cost down due to the bulk purchase of the raw or waste materials for the project. The logistics demands higher cost and may be minimized by considering above-mentioned solutions. These aspects have not been discussed in detail keeping the scope of the book in focus; however, using the appropriate LCA model such aspects may be included and the overall cost-effectiveness of the concrete being produced may be calculated.

In summary, the LCA of concrete is a powerful tool to ascertain several outcomes from the concrete-making process with and without waste utilization.

10 Future Research Areas in the Topic and Suggested Readings

10.1 FUTURE RESEARCH TOPICS IN THE AREA OF DEVELOPMENT OF SUSTAINABLE CONCRETE

The notion of sustainability raises awareness and consciousness regarding energy usage, mitigation of exploitation of the natural resources, and more importantly reusing the wastes generated during the industrial and consumer-oriented processes and productions. Construction activities have been one of the most demanding sectors by mankind for meeting the expanding needs of habitats and infrastructure. Conventional concrete, which is the most commonly utilized construction material on the one hand, has shown limitations of being environmentally friendly due to its limited recyclability and role in producing new materials. Therefore, researchers have made several efforts to reduce the use of natural materials in concrete and expend sincere efforts to develop concrete from wastes. We may consider past three decades as being the most active span wherein a large amount of wastes have been explored for such applications. At present, industrial and plastic wastes have become the most commonly utilized wastes in concrete making. Thanks to the modern scientific methods of microstructural investigations of the materials, civil engineers are capable of understanding the behavior of the modified composites including concrete in a better way and with more clarity. In this book, the author has provided a few interesting results by drawing our attention toward the potential of waste materials to possibly being mixed with conventional concrete, namely fly ash, silica fumes, slag, and waste plastic fibers. In this context, the author considers and suggests the following emerging areas to readers for their future works and investigations:

- Development of nano-scale waste materials by ball milling methods: Using the larger surface areas of conventional materials provides better intermolecular bonding and denser mix. This will increase the permeability resistance of the concrete composite and improve the service life of the structures.
- Detailed life cycle assessment of the concrete produced with waste material blending in binder as well as inert constituents for concrete: Partial to full replacement of the conventional materials from concrete has shown a few promising results for reduction of natural materials, which may possess the capacity to overcome the overuse of valuable natural resources. However,

DOI: 10.1201/9781032621340-10

from an energy consumption and eco-efficiency perspective, there is a need to obtain clarity on the advantages of making such modifications in concrete. This may be carried out by employing the LCA models on all aspects concerning the making of concrete using wastes. Use of free or licensed software may be a better option especially for inventory analysis.

- Development of methodology for pre-processing of the waste materials: One of the challenges of making sustainable concrete is the desired strength and durability performance of modified concrete. This is due to the limited information of the waste materials regarding their properties. The inherent properties of a waste material must be explored scientifically and for that a specific pre-treatment may be necessary. In most of the works, we can observe that the waste materials are not appropriately pre-processed or cleaned. This affects the desired or intended outcome of the modification process on the materials.
- Manufacturing of artificial aggregates: Unlike sand developed as manufactured sand and being used now in many applications, aggregates have not been successfully developed or are less explored. There is a need to explore the possibility of making aggregates using wastes and employ tests that may qualify the performance of the aggregates.

10.2   SUGGESTED READING

The author requests the readers to read more on the following topics and literature on the following areas and like to suggest a few documents and websites that may be of interest to all on the subject of the development of sustainable concrete composites.

Important e-books on the topic by CRC Press:

- *Green Building with Concrete: Sustainable Design and Construction*, Edited By Gajanan M. Sabnis Copyright 2016, Second Edition
- *Recycled Aggregate Concrete Technology and Properties*, By Natt Makul Copyright 2023
- *Sustainable Nano Materials for the Construction Industry*, By Ghasan Fahim Huseien, Kwok Wei Shah Copyright 2023
- *Life Cycle Assessment: Future Challenges*, By Surjya Narayana Pati Copyright 2023
- *Geopolymers as Sustainable Surface Concrete Repair Materials*, By Ghasan Fahim Huseien, Abdul Rahman Mohd Sam, Mahmood Md. Tahir Copyright 2023
- *Mine Waste Utilization*, By Ram Chandar Karra, Gayana B C, Shubhananda Rao P, Copyright 2022
- *Nanotechnology for Smart Concrete*, By Ghasan Fahim Huseien, Nur Hafizah A. Khalid, Jahangir Mirza, Copyright 2022
- *Concrete Recycling: Research and Practice*, Edited By Francois de Larrard, Horacio Colina, Copyright 2019

Useful web resources on the topic:

1. https://taylorandfrancis.com/search-results/?query=sustainable+building+
   materials&pg=&sort=&tab=
2. https://greenconcrete.berkeley.edu/
3. https://www.greenconcrete.info/

The readers are advised to refer to the standard codes and internationally recognized documents produced by the competitive authorities in the areas of concrete technology, sustainable concrete using cement-free binders, and similar literature using the journal papers published by the Tailor & Francis group by visiting the webpage taylorandfrancis.com. The author will be overjoyed to hear from the readers regarding the discussions and suggestions on any relevant topic of sustainable concrete technology.

Index

Note: **Bold** page numbers refer to tables and *italic* page numbers refer to figures.

For Product Safety Concerns and Information please contact our EU
representative GPSR@taylorandfrancis.com
Taylor & Francis Verlag GmbH, Kaufingerstraße 24, 80331 München, Germany